SpringerBriefs in Mathematics

SpringerBriefs present concise summaries of cutting-edge research and practical applications across a wide spectrum of fields. Featuring compact volumes of 50 to 125 pages, the series covers a range of content from professional to academic. Briefs are characterized by fast, global electronic dissemination, standard publishing contracts, standardized manuscript preparation and formatting guidelines, and expedited production schedules.

Typical topics might include:

- A timely report of state-of-the art techniques
- A bridge between new research results, as published in journal articles, and a contextual literature review
- A snapshot of a hot or emerging topic
- An in-depth case study
- A presentation of core concepts that students must understand in order to make independent contributions

SpringerBriefs in Mathematics showcases expositions in all areas of mathematics and applied mathematics. Manuscripts presenting new results or a single new result in a classical field, new field, or an emerging topic, applications, or bridges between new results and already published works, are encouraged. The series is intended for mathematicians and applied mathematicians.

Titles from this series are indexed by Web of Science, Mathematical Reviews, and zbMATH.

More information about this series at http://www.springer.com/series/10030

Kenkichi Iwasawa

Hecke's *L*-functions

Spring, 1964

Springer

Kenkichi Iwasawa
Princeton University
Princeton, NJ, USA

Foreword by
John Coates
Emmanuel College
Cambridge, UK

Masato Kurihara
Department of Mathematics
Keio University
Yokohama, Japan

ISSN 2191-8198 ISSN 2191-8201 (electronic)
SpringerBriefs in Mathematics
ISBN 978-981-13-9494-2 ISBN 978-981-13-9495-9 (eBook)
https://doi.org/10.1007/978-981-13-9495-9

This Springer imprint is published by the registered company Springer Nature Singapore Pte Ltd.
The registered company address is: 152 Beach Road, #21-01/04 Gateway East, Singapore 189721, Singapore

K. Iwasawa
Princeton, 1986

Now, $W(\chi) = \chi(\delta\varphi)W'(\chi;\delta,\varphi)$ by the definition. Hence it follows from Proposition 2, §3, II, that

$$|W(\chi)| = |\chi(\delta\varphi)||W'(\chi;\delta,\varphi)| = 1,$$
$$\overline{W(\chi)} = \overline{\chi}(\delta\varphi)W'(\overline{\chi};\delta,\varphi) = W(\overline{\chi}),$$

so that

$$W(\chi)W(\overline{\chi}) = 1.$$

On the other hand, if $\chi = 1$, then $W(1) = 1$. Hence we obtain from (1) that

$$\xi(s;\chi) = W(\chi)\left(W(\overline{\chi})\eta(s;\chi) + \eta(1-s;\overline{\chi}) + \frac{\varepsilon v}{s(s-1)}\right)$$
$$= W(\chi)\,\xi(1-s;\overline{\chi}),$$

namely,

$$\xi(s;\chi) = W(\chi)\,\xi(1-s;\overline{\chi}).$$

By Proposition 1, §4, we have

$$(2)\qquad L(s;\chi) = A(s;\chi)^{-1}\gamma(s;\chi)^{-1}\xi(s;\chi),$$

for $Re(s)>1$. Here $A(s;\chi)^{-1}$ is a function of the form e^{as+b}, $a, b = \text{const.}$ and $\gamma(s;\chi)^{-1}$ is a product of functions of the form $\Gamma(cs+d)^{-1}$. Since $\Gamma(s)^{-1}$ is holomorphic on the entire s-plane, $A(s;\chi)^{-1}\gamma(s;\chi)^{-1}$ is also holomorphic for arbitrary s. It then follows from the above that the L-function $L(s;\chi)$, which was originally defined for s with $Re(s)>1$, is a meromorphic function of s on the entire s-plane, satisfying (2) for arbitrary s. If $\chi \neq 1$, then $\xi(s;\chi)$ is an entire function of s. Hence $L(s;\chi)$ is also an entire function of s. Let $\chi \equiv 1$. $\zeta_F(s) = L(s;1)$ is holomorphic at $s \neq 0, 1$, and since

$$\gamma(s;1)^{-1} = \Gamma\left(\tfrac{s}{2}\right)^{-r_1}\Gamma(s)^{-r_2}, \qquad r_1+r_2>0,$$

has a zero at $s=0$, $\zeta_F(s)$ is still holomorphic at $s=0$. At $s=1$, $\xi(s;1)$ has a simple pole with residue $v = \mu_{\bar{J}_1}(\bar{J}_1)$, and

$$A(1,1)^{-1} = \frac{\pi^{\frac{r_1}{2}}}{\sqrt{d}}, \quad \gamma(1;1)^{-1} = \Gamma\left(\tfrac{1}{2}\right)^{-r_1} = \pi^{-\frac{r_1}{2}}.$$

Hence $\zeta_F(s)$ has a simple pole at $s=1$ with residue $\frac{v}{\sqrt{d}}$.

We now summarize our results as follows:

<u>Theorem 1.</u> The L-function $L(s;\chi)$ for a Hecke character χ of F, which was originally defined for s with $Re(s)>1$, is a meromorphic function of s on the entire s-plane. If $\chi \neq 1$, then $L(s;\chi)$ is holomorphic everywhere (i.e., an entire function of s). If $\chi \equiv 1$, then $\zeta_F(s) = L(s;1)$ has a unique simple pole at $s=1$ with residue $\frac{v}{\sqrt{d}}$, $v = \mu_{\bar{J}_1}(\bar{J}_1) < +\infty$. Let

$$\xi(s;\chi) = A(s;\chi)\gamma(s;\chi)L(s;\chi), \qquad s \in \mathbb{C},$$

with $A(s;\chi)$ and $\gamma(s;\chi)$ in §3. Then $\xi(s;\chi)$ satisfies the function equation

$$\xi(s;\chi) = W(\chi)\,\xi(1-s;\bar{\chi}), \qquad s \in \mathbb{C},$$

where $W(\chi)$ is a constant, depending only upon χ, such that

$$|W(\chi)| = 1, \quad \overline{W(\chi)} = W(\bar{\chi}).$$

Now, since $m_k, m_k' \geq 0$ and s_k are real,

$$\gamma(s;\chi) = \prod_{k=1}^{r} \Gamma\left(\frac{e_k}{2}\left(s + \frac{m_k + m_k'}{2} + is_k\right)\right)$$

does not vanish for $Re(s) > 1$. Hence $\xi(s;\chi) = A(s;\chi)\gamma(s;\chi)L(s;\chi)$ also has no zero for $Re(s) > 1$. By the functional equation, we then see that $\xi(s;\chi)$ has no zero for $Re(s) < 0$. Hence the zeros of $L(s;\chi)$ in the domain $Re(s) < 0$ are obtained from those of $\gamma(s;\chi)^{-1}$, and they can be easily determined because $\Gamma(s)^{-1}$ has only simple zeros at $s = 0, -1, -2, \cdots$. Thus all "non-trivial" zeros of $L(s;\chi)$ are in the critical strip $\{s \mid 0 \leq Re(s) \leq 1\}$. The generalized Riemann hypothesis states that all such zeros are on the straight line $Re(s) = \frac{1}{2}$.

In general, a locally compact group is compact if and only if it has a finite total Haar measure. In the above, we have shown that $\mu_{\bar{J}}(\bar{J}_1) < +\infty$. Hence:

Theorem 2. The group $\bar{J}_1 = J_1/F^*$ is a compact group.

We shall next show that the compactness of $\bar{J}_1$ implies two fundametal theorems on algebraic number fields, namely, the finiteness of class numbers and Dirichlet's unit theorem.

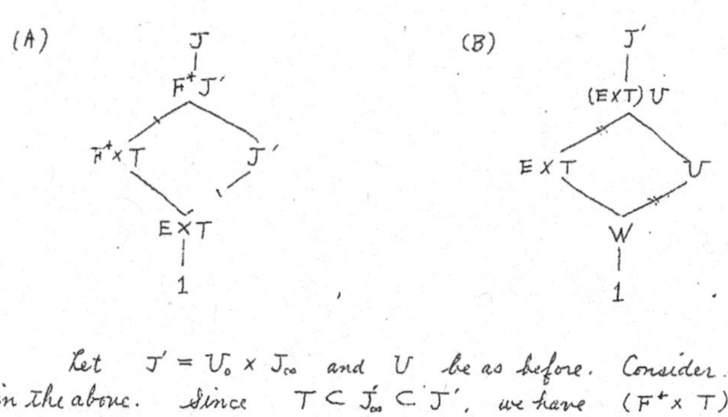

Let $J' = U_0 \times J_\infty$ and U be as before. Consider the diagrams in the above. Since $T \subset J_\infty' \subset J'$, we have $(F^* \times T)J' = F^*J'$, and $(F^* \times T) \cap J' = (F^* \cap J') \times T = E \times T$, where $E = F^* \cap J'$ is the group of units of F. Since $U \subset J_1$, $J = J_1 \times T$, $(E \times T) \cap U = E \cap U = W$.

Foreword

In 1964, Kenkichi Iwasawa gave a course of lectures at Princeton University on the adelic approach to Hecke's *L*-functions. The present book carefully reproduces Iwasawa's own beautifully handwritten notes used for the course, and follows faithfully his terminology.

Hecke's proof, for any number field, of the analytic continuation and functional equation of the abelian *L*-series, and more generally of his *L*-functions with Hecke characters, is of fundamental importance in algebraic number theory. Moreover, thanks to the theory of complex multiplication, it also establishes the analytic continuation and functional equation of the complex *L*-series of abelian varieties with complex multiplication. The modern adelic approach to Hecke's complicated classical theory was discovered independently by Iwasawa and Tate around 1950, and marked the beginning of the whole modern adelic approach to automorphic forms and *L*-series. While Tate's thesis at Princeton University in 1950 was finally published in 1967 in the volume *Algebraic Number Theory* edited by Cassels and Fröhlich, no detailed account of Iwasawa's work has previously appeared, beyond a very brief note in the Proceedings of the International Congress of Mathematicians in 1950, and a short letter to Dieudonné (in *Adv. Studies Pure Math.* 21, 1992). The lectures presented in this volume at last provide a detailed account of Iwasawa's work.

After two preliminary chapters on the basic local and global theory of number fields, and the theory of Haar measure on the group of idèles, Chap. 3 of the book establishes the basic expression, due to Iwasawa and Tate, for the complex *L*-series of the Hecke *L*-series $L(s, \chi)$ attached to an arbitrary Hecke character χ of a number field F as an integral over the idèle group J of F of an idelic theta function (see Sect. 3.4). Iwasawa then goes on to prove the analytic continuation and functional equation from this expression. Not only are his proofs both beautiful and fully detailed, but he also carefully explains the method in the simplest case of the Riemann zeta function. He then goes on to establish Dirichlet's formula for the residue at $s = 1$ of the complex zeta function of F, pointing out that an elegant argument involving the compactness of the idèle class group of F also gives a non-classical proof of the finiteness of the class number and the unit theorem.

In the final chapter, Iwasawa succinctly explains the link between the adelic approach and the classical theory. He then goes on to give detailed proofs of key classical results on the distribution of prime ideals, and on the class number formulae for cyclotomic fields.

This volume provides an elegant and detailed account of questions which are of seminal importance for modern number theory, and it covers material which is not treated as fully or as elegantly in other basic texts on algebraic number theory. We believe that it will provide an ideal text for future courses on this central part of number theory. Finally, we warmly thank Takahiro Kitajima for his accurate conversion of Iwasawa's handwritten notes into LaTeX, and Rei Otsuki for his help with proofreading.

Cambridge, UK — John Coates
Yokohama, Japan — Masato Kurihara
February 2019

Contents

Chapter 1
Algebraic Number Fields

Notation:

$\mathbf{Z}$ = the ring of integers, $0, \pm 1, \pm 2, \ldots$,
$\mathbf{Q}$ = the rational field,
$\mathbf{R}$ = the real field,
$\mathbf{C}$ = the complex field.

1.1 Ideals

A complex number ξ is called an algebraic integer if $\mathbf{Z}[\xi]$ is a finitely generated $\mathbf{Z}$-module; this condition is equivalent to the fact that $f(\xi) = 0$ for some polynomial $f(X) = X^m + a_1 X^{m-1} + \cdots + a_m, a_i \in \mathbf{Z}$. Let $\mathbf{A}$ be the set of all algebraic integers. Then $\mathbf{A}$ is a subring of $\mathbf{C}$, and $\mathbf{A} \cap \mathbf{Q} = \mathbf{Z}$.

Let F be a finite algebraic number field, i.e. $\mathbf{Q} \subset F \subset \mathbf{C}$, $[F : \mathbf{Q}] < \infty$. In the following, such a field will be simply called a number field. Let $\mathfrak{o} = \mathbf{A} \cap F$. Then $\mathfrak{o}(\neq 0)$ is a subring of F, and F is its quotient field. A finitely generated $\mathfrak{o}$-submodule of F is called an ideal (a fractional ideal) of F. Ideals of the ring $\mathfrak{o}$ (in the usual sense) are ideals of F; they are called integral ideals of F. Let $\mathfrak{I}$ denote the set of all non-zero ideals of F. For $\mathfrak{a}, \mathfrak{b} \in \mathfrak{I}$, $\mathfrak{a} + \mathfrak{b} (= (\mathfrak{a}, \mathfrak{b}))$ and $\mathfrak{a}\mathfrak{b}$ are defined as usual. They are again ideals in $\mathfrak{I}$.

A fundamental theorem states that $\mathfrak{I}$ is a free abelian group with respect to the above multiplication, and that the set of all non-zero prime ideals of $\mathfrak{o}$ forms a system of free generators of $\mathfrak{I}$. Namely, each ideal $\mathfrak{a}$ in $\mathfrak{I}$ can be uniquely written in the form

$$\mathfrak{a} = \prod_{\mathfrak{p}} \mathfrak{p}^{m_{\mathfrak{p}}}, \quad m_{\mathfrak{p}} \in \mathbf{Z} \tag{1.1}$$

where $m_{\mathfrak{p}} = 0$ for almost all $\mathfrak{p}$. Furthermore, $\mathfrak{a}$ is an integral ideal if and only if $m_{\mathfrak{p}} \geq 0$ for every $\mathfrak{p}$. The exponent $m_{\mathfrak{p}}$ in (1.1), which is uniquely determined by $\mathfrak{a}$, will be denoted by $\nu_{\mathfrak{p}}(\mathfrak{a})$.

K. Iwasawa, *Hecke's L-functions*, SpringerBriefs in Mathematics,
https://doi.org/10.1007/978-981-13-9495-9_1

For any $a \neq 0$ in F, let $(a) = a\mathfrak{o}$. Then (a) is an ideal in $\mathfrak{I}$; it is called the principal ideal generated by a. $(1) = \mathfrak{o}$ is of course the unit of $\mathfrak{I}$, and the set of all principal ideals forms a subgroup $\mathfrak{H}$ of $\mathfrak{I}$. $\mathfrak{I}/\mathfrak{H}$ is called the ideal class group of F. It is a finite group, and its order, $[\mathfrak{I} : \mathfrak{H}]$, is called the class number of F.

Let $\mathfrak{p}$ be a prime ideal of $\mathfrak{o}$. Then $\mathfrak{o}/\mathfrak{p}$ is a finite field. The number of elements in $\mathfrak{o}/\mathfrak{p}$, $[\mathfrak{o} : \mathfrak{p}]$, is called the norm of $\mathfrak{p}$, and is denoted by $N(\mathfrak{p})$. Let $\mathfrak{a}$ be any ideal in $\mathfrak{I}$, and let $\mathfrak{a}$ be written in the form (1.1). We define the norm of $\mathfrak{a}$ by

$$N(\mathfrak{a}) = \prod_{\mathfrak{p}} N(\mathfrak{p})^{m_{\mathfrak{p}}}.$$

Then $N(\mathfrak{ab}) = N(\mathfrak{a})N(\mathfrak{b})$ for any $\mathfrak{a}$, $\mathfrak{b}$ in $\mathfrak{I}$, and if $\mathfrak{a}$ is integral, then $N(\mathfrak{a}) = [\mathfrak{o} : \mathfrak{a}]$. If $\mathfrak{a} = a\mathfrak{o}$, then $N(\mathfrak{a}) = |N(a)|$ with $N(a)$ the norm of a.

An element $a \neq 0$ in F is called a unit of F if $(a) = \mathfrak{o}$; $a \neq 0$ is a unit if and only if both a and a^{-1} are in $\mathfrak{o}$. The set of all units of F forms a subgroup E of the multiplicative group F^* of F. Let W be the set of all roots of unity contained in F. Then W is a finite subgroup of E, and E/W is a free abelian group of rank $r_1 + r_2 - 1$ (see below).

Let $n = [F : \mathbf{Q}]$. Then there exist exactly n isomorphisms $\sigma_1, \ldots, \sigma_n$ from F into $\mathbf{C}$. Let $\sigma : F \to \mathbf{C}$ be one of them. Then $\overline{\sigma} : F \to \mathbf{C}$, defined by $\overline{\sigma}(a) = \overline{\sigma(a)}$ = the complex conjugate of $\sigma(a)$, is also an isomorphism. We call σ real or complex according as $\sigma = \overline{\sigma}$ or $\sigma \neq \overline{\sigma}$. We shall always fix the indices so that $\sigma_1, \ldots, \sigma_{r_1}$ are real, $\sigma_{r_1+1}, \ldots, \sigma_n$ are complex, and $\overline{\sigma_{r_1+i}} = \sigma_{r_1+r_2+i}$, $1 \leq i \leq r_2$, $n = r_1 + 2r_2$.

For each a in F, let $a^{(i)} = \sigma_i(a)$ for $i = 1, 2, \ldots, n$. Let $\alpha = (a_1, \ldots, a_n)$ be any ordered basis of F over $\mathbf{Q}$. Define an $n \times n$ matrix M_α by $M_\alpha = (a_i^{(j)})$. Then $M_\alpha {}^t M_\alpha = (T(a_i a_j))$, where T denotes the trace map from F to $\mathbf{Q}$. Put $D(\alpha) = \det M_\alpha {}^t M_\alpha = (\det M_\alpha)^2 = \det(T(a_i a_j))$. Then $D(\alpha)$ is a rational number, and $D(\alpha) \neq 0$. It follows that there exists a unique basis $\beta = (b_1, \ldots, b_n)$ such that

$$T(a_i b_j) = \delta_{ij}, \quad i, j = 1, \ldots, n,$$

where δ_{ij} is Kronecker's delta. The basis β is called the complementary basis to α. (α is then complementary to β.) The above equality is equivalent to $M_\alpha {}^t M_\beta = I$, namely,

$$M_\alpha = {}^t M_\beta^{-1}, \quad M_\beta = {}^t M_\alpha^{-1}.$$

Let $\mathfrak{a}$ be an ideal in $\mathfrak{I}$. Then $\mathfrak{a}$ has a basis $\alpha = (a_1, \ldots, a_n)$ over $\mathbf{Z}$, which is also a basis of F over $\mathbf{Q}$. For any such basis α of $\mathfrak{a}$, $D(\alpha)$ is always the same. Hence we denote it by $D(\mathfrak{a})$. In particular, we put $\Delta = D(\mathfrak{o})$, and call Δ the discriminant of F. Then $D(\mathfrak{a}) = \Delta N(\mathfrak{a})^2$ for any ideal $\mathfrak{a}$. Δ is a rational integer, and $d = |\Delta| > 1$ for $F \neq \mathbf{Q}$.

Let $\tilde{\mathfrak{a}}$ denote the set of all b in F such that $T(ab)$ is in $\mathbf{Z}$ for any a in an ideal $\mathfrak{a}$ in $\mathfrak{I}$. Then $\tilde{\mathfrak{a}}$ is also an ideal in $\mathfrak{I}$. Let β be the complementary basis to a basis α of $\mathfrak{a}$ over $\mathbf{Z}$. Then β is a basis of $\tilde{\mathfrak{a}}$ over $\mathbf{Z}$. Let $\mathfrak{d} = \tilde{\mathfrak{o}}^{-1}$. Then $\mathfrak{d}$ is an integral ideal, and $N(\mathfrak{d}) = |\Delta| = d$. By the definition, $\mathfrak{d}^{-1} = \tilde{\mathfrak{o}}$ is the largest ideal of F such that $T(\tilde{\mathfrak{o}}) \subset \mathbf{Z}$. $\mathfrak{d}$ is called the different of F. We have $\tilde{\mathfrak{a}} = \mathfrak{a}^{-1}\mathfrak{d}^{-1}$ for any ideal $\mathfrak{a}$.

1.2 Valuations (Absolute Values) and Prime Spots

Let F be any field. A map $\mathrm{v} : F \to \mathbf{R}$ is called a valuation of F if

(i) $\mathrm{v}(a) \geq 0$, $a \in F$; $\mathrm{v}(a) = 0 \Leftrightarrow a = 0$,
(ii) $\mathrm{v}(ab) = \mathrm{v}(a)\mathrm{v}(b)$, $a, b \in F$,
(iii) $\mathrm{v}(a + b) \leq C \max(\mathrm{v}(a), \mathrm{v}(b))$, $a, b \in F$, with some constant $C > 0$.

Let $a \neq 0$, $b = 0$ in (iii). Then $C \geq 1$. Let $\|\mathrm{v}\| =$ inf. of all C satisfying (iii). Then $\|\mathrm{v}\| \geq 1$, and iii) is still satisfied with $C = \|\mathrm{v}\|$. The valuation v is called archimedean or non-archimedean according as $\|\mathrm{v}\| > 1$ or $\|\mathrm{v}\| = 1$.

Let v be a valuation of F. Then, for any real $\alpha > 0$, $\mathrm{v}' = \mathrm{v}^{\alpha}$ is also a valuation of F. Two valuations of F are called equivalent if one is a power of the other: $\mathrm{v}' = \mathrm{v}^{\alpha}$, $\alpha > 0$. Each class of equivalent valuations of F is called a prime spot[1] (or spot, or prime divisor) of F. Two equivalent valuations of F are either both archimedean or both non-archimedean. Hence we call a prime spot of F archimedean or non-archimedean according as it consists of archimedean or non-archimedean valuations.

Let $\mathrm{v}(0) = 0$, $\mathrm{v}(a) = 1$ for $a \neq 0$ in F. Then v is a valuation of F, and it alone constitutes a non-archimedean prime spot of F, called the trivial prime spot of F.

Let v be a valuation of F. For each $\alpha > 0$, let U_{α} denote the subset of all (a, b) in $F \times F$ such that $\mathrm{v}(a - b) < \alpha$. Then $\{U_{\alpha}\}$ defines a uniform structure on F, and F is a topological field relative to the Hausdorff topology defined by that uniform structure. Two valuations of F are equivalent if and only if they define the same uniform structure on F. Hence each prime spot P of F determines a uniform structure on F. Let F_P denote the completion of F relative to that uniform structure. Then we can make F_P into a topological field in a natural way so that F is a subfield of F_P everywhere dense in F_P. Each valuation v in P can be uniquely extended by continuity to a valuation v' on F_P, and these valuations v' form a prime spot on F_P. F_P is called the P-completion of F.

Let K be a subfield of F. The restriction of a valuation $\mathrm{v} : F \to \mathbf{R}$ on the subfield K defines a valuation $w : K \to \mathbf{R}$. Let P be a prime spot of F. Then the restrictions of valuations in P on the subfield K fulfill a prime spot Q of K. We call Q the restriction of P on K, and P an extension of Q on F. Q is archimedean or non-archimedean according as P is archimedean or non-archimedean. The Q-completion K_Q of K may be identified with the closure of K in the P-completion F_P of F so that F_P is an extension field of K_Q.

Let $P_1, \ldots, P_m$ be a finite set of prime spots of F. Let v_i be a valuation in P_i, and let a_i be an arbitrary element of F_{P_i}, $i = 1, \ldots, m$. Then for any small $\varepsilon > 0$, there exists an element a in F such that

[1] The modern terminology of a "prime spot" is a "place".

$$\mathrm{v}_i(a - a_i) < \varepsilon, \quad i = 1, \ldots, m.$$

This is called the approximation theorem.

Now let F be a number field. We denote by S (resp. S_0, S_∞) the set of all non-trivial spots (resp. non-archimedean spots, archimedean spots) of F. Spots in S_0 are also called finite spots, and spots in S_∞ infinite spots. Let $\sigma_1, \ldots, \sigma_n$ be as in Sect. 1.1, $n = [F : \mathbf{Q}]$. Let

$$\mathrm{v}_{\infty,i}(a) = \begin{cases} |\sigma_i(a)| & (1 \leq i \leq r_1), \\ |\sigma_i(a)|^2 = |\sigma_i(a)\sigma_{i+r_2}(a)| & (r_1 + 1 \leq i \leq r_1 + r_2) \end{cases}$$

for any a in F. Then $\mathrm{v}_{\infty,i}$ are archimedean valuations of F, inequivalent to each other. Let $P_{\infty,i}$ denote the prime spot of $\mathrm{v}_{\infty,i}$. Then

$$S_\infty = \{P_{\infty,1}, \ldots, P_{\infty,r_1+r_2}\}.$$

$P_{\infty,i}$ is called real or complex according as σ_i is real or complex. σ_i can be extended to an isomorphism of $F_{P_{\infty,i}}$ *onto* $\mathbf{R}$ or $\mathbf{C}$ according as σ_i is real or complex. The valuation $\mathrm{v}_{\infty,i}$ will be also denoted by $\mathrm{v}_{P_{\infty,i}}$.

Let $\mathfrak{p}$ be a prime ideal of $\mathfrak{o}$. Define $\mathrm{v}_\mathfrak{p} : F \to \mathbf{R}$ by

$$\mathrm{v}_\mathfrak{p}(a) = N(\mathfrak{p})^{-\nu_\mathfrak{p}(a)}, \quad a \in F,$$

where $\nu_\mathfrak{p}(a) = +\infty$ for $a = 0$, and $\nu_\mathfrak{p}(a) = \nu_\mathfrak{p}((a))$ for $a \neq 0$. Then $\mathrm{v}_\mathfrak{p}$ is a non-trivial non-archimedean valuation of F, and we denote the prime spot of $\mathrm{v}_\mathfrak{p}$ by $P_\mathfrak{p}$. For $\mathfrak{p} \neq \mathfrak{p}'$, we have $P_\mathfrak{p} \neq P_{\mathfrak{p}'}$, and when $\mathfrak{p}$ ranges over all prime ideals of $\mathfrak{o}$,

$$S_0 = \{P_\mathfrak{p}\}.$$

Thus there is a one-one correspondence $P \leftrightarrow \mathfrak{p}$ between the set S_0 and the set of all prime ideals of $\mathfrak{o}$. The prime ideal corresponding to a prime spot P will be denoted by $\mathfrak{p}_P$. We also write v_P, ν_P, and $F_\mathfrak{p}$ for $\mathrm{v}_\mathfrak{p}$, $\nu_\mathfrak{p}$ and F_P respectively if $P \leftrightarrow \mathfrak{p}$.

Now, v_P is defined for each P in S. For any $a \neq 0$ in F, $\mathrm{v}_P(a) = 1$ for almost all P in S, and

$$\prod_{P \in S} \mathrm{v}_P(a) = 1.$$

This is called the product formula.

Let $P_1, \ldots, P_m$ be any finite subset of S_0. Suppose that for each i with $1 \leq i \leq m$, an element a_i in F_{P_i} and an integer $m_i \geq 0$ are given. Then there exists an element a in F such that

$$\nu_{P_i}(a - a_i) \geq m_i, \qquad i = 1, \ldots, m,$$
$$\nu_P(a) \geq 0, \qquad P \in S_0, \; P \neq P_1, \ldots, P_m.$$

This is called the strong approximation theorem.

The restriction of a non-trivial prime spot of F on the subfield $\mathbf{Q}$ is a non-trivial prime spot of $\mathbf{Q}$. Hence prime spots in S are obtained by extending non-trivial prime spots of $\mathbf{Q}$ on F.

The rational field $\mathbf{Q}$ has a unique infinite prime spot P_∞ ($r_1 = 1, r_2 = 0$) and the P_∞-completion of $\mathbf{Q}$ is $\mathbf{R}$. The extensions of P_∞ on F give us all infinite spots of F, $P_{\infty,i}$, $1 \leq i \leq r_1 + r_2$.

A prime number p defines a prime ideal $p\mathbf{Z}$ of $\mathbf{Z}$, and hence a finite prime spot P_p of $\mathbf{Q}$. The P_p, $p = 2, 3, 5, 7, 11, \ldots$, gives us all non-trivial finite spots of $\mathbf{Q}$. The P_p-completion of $\mathbf{Q}$ is the p-adic number field $\mathbf{Q}_p$. Let $(p) = p\mathfrak{o} = \prod_{i=1}^{g} \mathfrak{p}_i^{e_i}$ be the prime ideal decomposition of (p), $\mathfrak{p}_i \neq \mathfrak{p}_j$ $(i \neq j)$, $e_i \geq 1$. Then $P_{\mathfrak{p}_i}$, $i = 1, \ldots, g$, give us all extensions of P_p on F.

Let Q be a non-trivial prime spot of $\mathbf{Q}$, finite or infinite, and let $P_1, \ldots, P_g$ be the extensions of Q on F. Then there exists an isomorphism of $\mathbf{Q}_Q$-algebras

$$F \otimes_{\mathbf{Q}} \mathbf{Q}_Q \cong F_{P_1} \oplus \cdots \oplus F_{P_g}.$$

Let P be any one of $P_1, \ldots, P_g$. Let $n_P = [F_P : \mathbf{Q}_Q]$, and let T_P and N_P denote the trace and the norm map from F_P to $\mathbf{Q}_Q$. Then it follows from the above that

$$n = \sum_{i=1}^{g} n_{P_i}, \quad T(a) = \sum_{i=1}^{g} T_{P_i}(a), \quad N(a) = \prod_{i=1}^{g} N_{P_i}(a)$$

for any a in F.

Let $\mathfrak{p}$ be a prime ideal of $\mathfrak{o}$. The extensions of the functions $\mathrm{v}_\mathfrak{p}$ and $\nu_\mathfrak{p}$ or $F_\mathfrak{p}$ ($= F_P$, $P = P_\mathfrak{p}$) will be denoted again by the same letters. Let

$$\mathfrak{o}_\mathfrak{p} = \{a \mid a \in F_\mathfrak{p}, \; \nu_\mathfrak{p}(a) \geq 0\}$$
$$\mathfrak{m}_\mathfrak{p} = \{a \mid a \in F_\mathfrak{p}, \; \nu_\mathfrak{p}(a) > 0\}.$$

Then $\mathfrak{o}_\mathfrak{p}$ is a subring of $F_\mathfrak{p}$, $\mathfrak{m}_\mathfrak{p}$ is the unique maximal ideal of $\mathfrak{o}_\mathfrak{p}$, $\mathfrak{o}_\mathfrak{p} = \mathfrak{o} + \mathfrak{m}_\mathfrak{p}$, $\mathfrak{p} = \mathfrak{o} \cap \mathfrak{m}_\mathfrak{p}$ so that $\mathfrak{o}_\mathfrak{p}/\mathfrak{m}_\mathfrak{p} \cong \mathfrak{o}/\mathfrak{p}$. Elements of $\mathfrak{o}_\mathfrak{p}$ are called $\mathfrak{p}$-adic integers. $F_\mathfrak{p}$ is a locally compact topological field, and both $\mathfrak{o}_\mathfrak{p}$ and $\mathfrak{m}_\mathfrak{p}$ are open and compact in $F_\mathfrak{p}$. A finitely generated $\mathfrak{o}_\mathfrak{p}$-module in $F_\mathfrak{p}$ is called an ideal of $F_\mathfrak{p}$. Every non-zero ideal of $F_\mathfrak{p}$ is principal and is a power of $\mathfrak{m}_\mathfrak{p}$; $\mathfrak{a} = a\mathfrak{o}_\mathfrak{p} = \mathfrak{m}_\mathfrak{p}^s$, $s = \nu_\mathfrak{p}(a)$. For such $\mathfrak{a}$, let $\tilde{\mathfrak{a}}$ be the set of all b in $F_\mathfrak{p}$ such that $T_\mathfrak{p}(ab)$ is in $\mathbf{Z}_p$ for any a in $\mathfrak{a}$; here p is the prime number such that $Q = P_p$ is the restriction of P on $\mathbf{Q}$, $T_\mathfrak{p} = T_P$ is the trace from F_P to $\mathbf{Q}_Q = \mathbf{Q}_p$, and $\mathbf{Z}_p$ is the ring of p-adic integers in $\mathbf{Q}_p$. Then $\tilde{\mathfrak{a}}$ is also a non-zero ideal of $F_\mathfrak{p}$. Let $\mathfrak{d}_\mathfrak{p} = \tilde{\mathfrak{o}}_\mathfrak{p}^{-1}$. Then $\mathfrak{d}_\mathfrak{p} \subset \mathfrak{o}_\mathfrak{p}$. We call $\mathfrak{d}_\mathfrak{p}$ the different of $F_\mathfrak{p}$. $\tilde{\mathfrak{a}} = \mathfrak{a}^{-1}\mathfrak{d}_\mathfrak{p}^{-1}$, and $\mathfrak{d}_\mathfrak{p} = \mathfrak{o}_\mathfrak{p}$ if and only if $e_\mathfrak{p}$ $(= e_P) = 1$.

In general, let $\mathfrak{a}$ be an ideal of F, $\mathfrak{a} \neq \{0\}$, and let $\mathfrak{a} = \prod \mathfrak{p}^{s_\mathfrak{p}}$, $s_\mathfrak{p} = \nu_\mathfrak{p}(\mathfrak{a})$. Let $\mathfrak{a}_\mathfrak{p}$ be the closure of $\mathfrak{a}$ in $F_\mathfrak{p}$ ($\mathfrak{a} \subset F \subset F_\mathfrak{p}$). Then $\mathfrak{a}_\mathfrak{p} = \mathfrak{m}_\mathfrak{p}^{s_\mathfrak{p}} = \{a \mid a \in F_\mathfrak{p},\ \nu_\mathfrak{p}(a) \geq s_\mathfrak{p} = \nu_\mathfrak{p}(\mathfrak{a})\}$, and

$$\mathfrak{a} = \bigcap_\mathfrak{p} (F \cap \mathfrak{a}_\mathfrak{p}).$$

Conversely, let $\{s_\mathfrak{p}\}$ be a system of integers such that $s_\mathfrak{p} = 0$ for almost all $\mathfrak{p}$. Let $\mathfrak{a}_\mathfrak{p} = \mathfrak{m}_\mathfrak{p}^{s_\mathfrak{p}}$, and let $\mathfrak{a} = \bigcap_\mathfrak{p} (F \cap \mathfrak{a}_\mathfrak{p}) = \{a \mid a \in F,\ \nu_\mathfrak{p}(a) \geq s_\mathfrak{p} \text{ for every } \mathfrak{p}\}$. Then $\mathfrak{a}$ is an ideal of F, $\mathfrak{a} \neq \{0\}$ and

$$\mathfrak{a} = \prod_\mathfrak{p} \mathfrak{p}^{s_\mathfrak{p}}.$$

Let $\mathfrak{d}$ be the different of F. Then the closure of $\mathfrak{d}$ in $F_\mathfrak{p}$ is nothing but the different $\mathfrak{d}_\mathfrak{p}$ of $F_\mathfrak{p}$ defined in the above, and if $\mathfrak{d} = \prod \mathfrak{p}^{t_\mathfrak{p}}$, $t_\mathfrak{p} = \nu_\mathfrak{p}(\mathfrak{d})$, then $\mathfrak{d}_\mathfrak{p} = \mathfrak{m}^{t_\mathfrak{p}}$, and

$$\mathfrak{d} = \bigcap_\mathfrak{p} (F \cap \mathfrak{d}_\mathfrak{p}).$$

Since $t_\mathfrak{p} = \nu_\mathfrak{p}(\mathfrak{d}) = 0$ for almost all $\mathfrak{p}$, we have $\mathfrak{d}_\mathfrak{p} = \mathfrak{o}_\mathfrak{p}$ for almost all $\mathfrak{p}$ namely, $e_\mathfrak{p}\ (= e_P) = 1$ for almost all $\mathfrak{p}$ (almost all P in S).

The multiplicative group $F_\mathfrak{p}^*$ of the locally compact field $F_\mathfrak{p}$ is a locally compact abelian group in the natural topology. The groups $1 + \mathfrak{m}_\mathfrak{p}^s$, $s \geq 1$, are open, compact subgroups of $F_\mathfrak{p}^*$, and form a base of neighborhoods of 1 in $F_\mathfrak{p}^*$. Let

$$U_\mathfrak{p}\ (= U_P) = \{a \mid a \in F_\mathfrak{p},\ \nu_\mathfrak{p}(a) = 0\} = \{a \mid a \in F_\mathfrak{p},\ \mathrm{v}_\mathfrak{p}(a) = 1\}.$$

Then $U_\mathfrak{p}$ is also an open, compact subgroup of $F_\mathfrak{p}^*$. $U_\mathfrak{p}$ contains $1 + \mathfrak{m}_\mathfrak{p}$, and $[U_\mathfrak{p} : 1 + \mathfrak{m}_\mathfrak{p}] = N(\mathfrak{p}) - 1$. $U_\mathfrak{p}$ is the kernel of $\nu_\mathfrak{p} : F_\mathfrak{p}^* \to \mathbf{Z}$ so that $F_\mathfrak{p}^*/U_\mathfrak{p} \cong \mathbf{Z}$.

Let $P = P_{\infty,i}$, $1 \leq i \leq r_1 + r_2$. Then we have the topological isomorphism

$$\sigma_i : F_P \to \mathbf{R} \text{ or } \mathbf{C}.$$

Hence F_P is again a locally compact field, and F_P^*, the multiplicative group of F_P, is a locally compact abelian group. Let

$$U_P = \{a \mid a \in F_P,\ \mathrm{v}_P(a) = 1\}.$$

Then

$$U_P \cong \begin{cases} \{\pm 1\}, & \text{if } \sigma_i \text{ is real,} \\ \mathbf{C}_1 = \{z \mid z \in \mathbf{C},\ |z| = 1\} & \text{if } \sigma_i \text{ is complex.} \end{cases}$$

Hence U_P is always a compact subgroup of F_P^*.

Remark 1.1 Let μ be a Haar measure of the additive group of F_P. Then for any measurable subset A of F_P, we have

$$\mu(aA) = \mathrm{v}_P(a)\mu(A), \quad a \in F_P.$$

This gives us a uniform, invariant definition of v_P.

Chapter 2
Idèles

2.1 Adèles and Idèles

Let F be a number field, and let S, S_0, S_∞, F_P etc. be as before. Put

$$R^* = \prod_{P \in S} F_P.$$

R^* is the set of all vectors $(\ldots, a_P, \ldots)$ where the P-components a_P are taken arbitrarily from F_P. R^* is a ring in the obvious manner; it is commutative and it has the identity $1 = (\ldots, 1, 1, 1, \ldots)$. Let a be any element of F. Then R^* contains the vector $(\ldots, a, a, a, \ldots)$. The map $a \mapsto (\ldots, a, a, a, \ldots)$ is obviously an injection of F into R^*. Hence we may identify a with $(\ldots, a, a, a, \ldots)$, and consider F as a subfield of R^*.

A vector $\alpha = (a_P)$ in R^* is called an adèle of F if $\mathrm{v}_P(a_P) \leq 1$ for almost all P in S, namely, if a_P is in $\mathfrak{o}_P$ for almost all P in S_0. The set of all adèles of F forms a subring R of R^*, called the adèle ring of F. If a is in F, $a \neq 0$, then $\mathrm{v}_P(a) = 1$ for almost all P in S. Hence a is in R. F is therefore a subfield of R.

Let

$$R' = \prod_{P \in S_0} \mathfrak{o}_P \times \prod_{P \in S_\infty} F_P.$$

Then R' is a subring of R. R' is a topological ring relative to the topology which is the product of the natural topologies on $\mathfrak{o}_P$, $P \in S_0$, and on F_P, $P \in S_\infty$. Since $\mathfrak{o}_P$'s are compact and F_P's are finite in number, R' is locally compact. Let $\{O\}$ be the family of all neighborhoods of 0 in R'. We define a topology on R by taking $\{\alpha + O\}$ as a base of neighborhoods of α in R. Then R becomes a (Hausdorff) topological ring, and R' an open subring of R. It follows in particular that R is also locally compact.

Problem 2.1 Is it possible to make R^* a topological ring in a similar way?

Proposition 2.1 *F is discrete in R.*

K. Iwasawa, *Hecke's L-functions*, SpringerBriefs in Mathematics,
https://doi.org/10.1007/978-981-13-9495-9_2

Proof Let $O = \prod \mathfrak{o}_P \times O_1 \times \cdots \times O_r$, $r = r_1 + r_2$, where $O_i = \{x \mid x \in F_{P_{\infty,i}}, \nu_{P_{\infty,i}}(x) < 1\}$. Then O is a neighborhood of 0 in R', and hence in R. Let a be in $F \cap O$. Then $a \in \bigcap_P (F \cap \mathfrak{o}_P) = \mathfrak{o}$ (cf Sect. 1.2). Hence $N(a) = \prod_{i=1}^{n} \sigma_i(a)$ is an algebraic integer in $\mathbf{Q}$, namely, a rational integer. On the other hand, $|N(a)| = \prod_{i=1}^{n} |\sigma_i(a)| = \prod_{i=1}^{r} \nu_{P_{\infty,i}}(a) < 1$. Hence $a = 0$. It follows that $F \cap O = \{0\}$ and that F is discrete in R.

Remark 2.1 R/F is compact.

An invertible element of the ring R is called an idèle of F. The set of all idèles of F forms a multiplicative group J, called the idèle group of F. It is easy to see that $\alpha = (a_P)$ is an idèle of F if and only if $a_P \neq 0$ for every P in S and $\nu_P(a_P) = 1$ for almost all P in S. Clearly the multiplicative group F^* of the field F is a subgroup of J.

Let

$$J' = \prod_{P \in S_0} U_P \times \prod_{P \in S_\infty} F_P^*.$$

J' is a subgroup of J. Since U_P is compact and F_P^* is locally compact, J' is a locally compact group as product of U_P, $P \in S_0$, and F_P^*, $P \in S_\infty$. We then make J into a locally compact group so that J' is an open subgroup of J (see the above argument for R' and R). For any subset A of J, the topology on A induced by that of R will be called the adèle topology of A, and the topology on A induced by that of J will be called the idèle topology of A. It then follows from the definition that the adèle topology and the idèle topology of J' are the same. Let O' be a neighborhood of 1 in J in the adèle topology. Then $O' = J \cap O$ for some neighborhood O of 1 in R. Let $O'' = J' \cap O$. Then O'' is a neighborhood of 1 in J' in the adèle topology, and hence in the idèle topology. Therefore O'' is also a neighborhood of 1 in J in the idèle topology of J. As $O'' \subset O'$, we see that the idèle topology of J is at least as strong as the adèle topology of J.

Proposition 2.2 *F^* is discrete in J (in the idèle topology of J).*

Proof F is discrete in R by Proposition 2.1. Hence $F^* = F \cap J$ is discrete in J in the adèle topology of J. Hence F^* is also discrete in J in the idèle topology of J because the latter is at least as strong as the former.

Problem 2.2 (i) For each neighborhood O of 1 in R, let O' denote the set of all α in J such that both α and α^{-1} are in O. Prove that $\{O'\}$ is a base of neighborhoods of 1 in J in the idèle topology of J.

(ii) Prove that the two topologies of J are different.

(iii) Let P be fixed. Prove that the map $R \to F_P$ (resp. $J \to F_P^*$) defined by $\alpha = (\ldots, a_P, \ldots) \mapsto a_P$ is continuous (resp. in the idèle topology of J).

Let

$$U = \prod_{P \in S} U_P.$$

Since U_P is always compact, U is a compact subgroup of J', and hence of J. Let

$$J_0 = \{\alpha \mid \alpha = (a_P) \in J,\ a_P = 1 \text{ for all } P \in S_\infty\},$$
$$J_\infty = \{\alpha \mid \alpha = (a_P) \in J,\ a_P = 1 \text{ for all } P \in S_0\}.$$

Then J_0 and J_∞ are closed subgroups of J such that

$$J = J_0 \times J_\infty.$$

Put

$$U_0 = U \cap J_0,\ U_\infty = U \cap J_\infty.$$

Then

$$U = U_0 \times U_\infty,\ J' = U_0 \times J_\infty,\ J/J' = J_0/U_0.$$

We define the volume $V(\alpha)$ of an idèle $\alpha = (a_P)$ by

$$V(\alpha) = \prod_P \mathrm{v}_P(a_P).$$

Since $\mathrm{v}_P(a_P) = 1$ for almost all P's, $V(\alpha)$ is well defined. It is easy to see that

$$V : J \to \mathbf{R}^+$$

is a continuous homomorphism; here $\mathbf{R}^+$ denotes the multiplicative group of positive real numbers. Let

$$\begin{aligned} J_1 &= \ker(V) \\ &= \{\alpha \mid \alpha \in J,\ V(\alpha) = 1\}. \end{aligned}$$

Clearly J_1 is a closed subgroup of J.

For each x in $\mathbf{R}^+$, let τ_x denote the vector (a_P) such that $a_P = 1$ for every P in S_0, and $a_{P_{\infty,i}} = \sigma_i^{-1}(x^{\frac{1}{n}})$ for $P_{\infty,i}$ in S_∞. Then τ_x is an idèle, and the set of all such τ_x, $x \in \mathbf{R}^+$, forms a subgroup T of J. Since $V(\tau_x) = x$, the maps $x \mapsto \tau_x \mapsto x = V(\tau_x)$ define a topological isomorphism

$$T \cong \mathbf{R}^+.$$

Furthermore, T is closed in J, and

$$J = J_1 \times T.$$

The product formula for F states that $V(a) = \prod v_P(a) = 1$ for any a in F^*. Hence

$$F^* \subset J_1,$$

and $F^*T = F^* \times T$ is a closed subgroup of J. Let

$$\overline{J} = J/F^*, \ \overline{J_1} = J_1/F^*, \ \overline{T} = F^*T/F^*.$$

Then

$$\overline{J} = \overline{J_1} \times \overline{T}, \ \overline{T} \cong T \cong \mathbf{R}^+,$$

topologically.

For $\alpha = (a_P)$ in J, let

$$\iota(\alpha) = \prod_{P \in S_0} \mathfrak{p}_P^{\nu_P(a_P)}.$$

Since $\nu_P(a_P) = 0$ for almost all P in S_0, the product is well defined, and $\iota(\alpha)$ is an ideal of F. The map

$$\iota : J \to \mathfrak{I} = \text{the ideal group of } F$$

is a surjective homomorphism, and its kernel is J'. The restriction of ι on J_0 is also surjective, and its kernel is $U_0 = J' \cap J_0$. Hence ι (and its restriction on J_0) induces

$$J/J' = J_0/U_0 \cong \mathfrak{I}.$$

Let $\alpha = (a_P)$ be in J_0. Then

$$V(\alpha) = \prod_{P \in S_0} v_P(a_P) = \prod_{S_0} N(\mathfrak{p}_P)^{-\nu_P(a_P)} = N\left(\prod \mathfrak{p}_P^{\nu_P(a_P)}\right)^{-1} = N(\iota(\alpha))^{-1},$$

namely,

$$V(\alpha) = N(\iota(\alpha))^{-1}, \ \alpha \in J_0.$$

A similar computation shows that

$$\iota(a) = (a), \ a \in F^*.$$

Hence

$$\iota(F^*) = \mathfrak{H} = \text{the group of principal ideals of } F$$

and we have

$$J/F^*J' \cong \mathfrak{I}/\mathfrak{H} = \text{the ideal class group of } F.$$

Let E denote as before the group of units in F. An element a in F^* is a unit if and only if $(a) = \mathfrak{o}$, namely, $\iota(a) = 1$. Hence

$$E = F^* \cap J'.$$

Let W be the group of roots of unity in F. Let ζ be in W, and let $\zeta^m = 1$. Then $\mathrm{v}_P(\zeta)^m = \mathrm{v}_P(\zeta^m) = \mathrm{v}_P(1) = 1$, and hence $\mathrm{v}_P(\zeta) = 1$ for every P. Therefore ζ is in $F^* \cap U$. However, since F^* is discrete and U is compact, $F^* \cap U$ must be finite. Hence each ζ in $F^* \cap U$ has a finite order: $\zeta^m = 1$, $m \geq 1$. It follows that

$$W = F^* \cap U.$$

We have shown at the same time that W is a finite group.

Remark 2.2 In general, we can define adèles for any algebraic variety V defined over a number field F. We shall next explain such adèles in the case where V is an affine algebraic group. A subgroup G of $\mathrm{SL}(m; F)$, $m \geq 1$, is called algebraic if there exists a set of polynomials $f_1, \ldots, f_s$ in $F[X_{11}, \ldots, X_{ij}, \ldots, X_{mm}]$ such that G consists of all $m \times m$ matrices $A = (a_{ij})$ over F, satisfying $f_k(\cdots, a_{ij}, \ldots) = 0$, for $k = 1, \ldots, s$. It is known in the theory of algebraic groups that an affine algebraic group defined over F is always isomorphic to such a subgroup of $\mathrm{SL}(m; F)$ for some $m \geq 1$.

Let $M(m; R)$ denote the ring of all $n \times n$ matrices over the adèle ring R of F; it is a locally compact ring in the obvious manner. A matrix $A = (\alpha_{ij})$ in $M(m; R)$ is called an adèle of G if $f_k(\ldots, \alpha_{ij}, \ldots) = 0$ for $k = 1, \ldots, s$. The set of all such adèles of G forms a multiplicative locally compact group $\widetilde{G}$ relative to the topology induced by that of $M(m; R)$. We call $\widetilde{G}$ the adèle group of G. The group G itself is then a discrete subgroup of $\widetilde{G}$.

Example 2.1 1. $G = \mathrm{SL}(m; F)$. In this case, $\widetilde{G} = \mathrm{SL}(m; R)$.

2. Let G be the group of all matrices $\begin{pmatrix} 1 & a \\ 0 & 1 \end{pmatrix}$ with a in F. Then $\widetilde{G}$ is the group of all matrices $\begin{pmatrix} 1 & \alpha \\ 0 & 1 \end{pmatrix}$ with arbitrary α in R, and $\widetilde{G} \cong R$ topologically.

3. Let G be the group of all matrices $\begin{pmatrix} a & 0 \\ 0 & b \end{pmatrix}$ with a, b in F, $ab = 1$. Then $\widetilde{G}$ is the group of all matrices $\begin{pmatrix} \alpha & 0 \\ 0 & \beta \end{pmatrix}$ with α, β in R, $\alpha\beta = 1$. We see immediately (use Problem) that $\begin{pmatrix} \alpha & 0 \\ 0 & \beta \end{pmatrix} \mapsto \alpha$ is a topological isomorphism of $\widetilde{G}$ onto J: $\widetilde{G} \cong J$.

2.2 Characters of R and J

A character of a topological group G is a continuous homomorphism $\chi : G \to \mathbf{C}_1$. We shall next consider characters of (the additive group of) R and J.

The following lemmas are easy to prove.

Lemma 2.1 *For any x in $\mathbf{Q}_p$, p = prime, there exists a unique rational number r such that (i) the denominator of r is a power of p, (ii) $0 \le r < 1$, (iii) $x \equiv r \bmod \mathbf{Z}_p$.*

We shall denote such an r by $\langle x \rangle_p$.

Lemma 2.2 *(i) $\langle x \rangle_p = 0$ $(x \in \mathbf{Q}_p) \Leftrightarrow x \in \mathbf{Z}_p$,*
(ii) $\langle x + y \rangle_p \equiv \langle x \rangle_p + \langle y \rangle_p \bmod \mathbf{Z}$, for any $x, y \in \mathbf{Q}_p$,
(iii) Let x be a rational number. Then $\langle x \rangle_p = 0$ for almost all primes p, and

$$\sum_p \langle x \rangle_p \equiv x \bmod \mathbf{Z}.$$

For each P in S, we shall next define a character ω_P of F_P, the additive group of F_P. Let P be infinite. Then put

$$\omega_P(a) = \exp\left(-2\pi i T_P(a)\right), \ a \in F_P,$$

where $\exp(z) = e^z$ and $T_P : F_P \to \mathbf{R}$ is, as before, the trace map. ω_P is obviously a character of F_P. Since (cf. Sect. 1.2)

$$T(a) = \sum_{P \in S_\infty} T_P(a), \ a \in F,$$

we obtain

$$\prod_{P \in S_\infty} \omega_P(a) = \exp(-2\pi i T(a)), \ a \in F.$$

Now, let P be finite, and let $p(= P_p)$ be the restriction of P on $\mathbf{Q}$. Denoting again by $T_P : F_P \to \mathbf{Q}_P$ the trace map, we put

$$\omega_P(a) = \exp(2\pi i \langle T_P(a) \rangle_p), \ a \in F_P.$$

Using Lemma 2.2, we see that $\omega_P(a + b) = \omega_P(a)\omega_P(b)$, $a, b \in F_P$. Let a be in $\mathfrak{d}_P^{-1} = \widetilde{\mathfrak{o}_P}$. Then $T_P(a)$ is in $\mathbf{Z}_p$ so that $\langle T_P(a) \rangle_p = 0$ and $\omega_P(a) = 1$. In particular, $\omega_P(a) = 1$ for any a in $\mathfrak{o}_P$. Hence ω_P is continuous. Therefore it is a character of F_P. When P ranges over all extensions of p on F, we have (cf. Sect. 1.2)

$$T(a) = \sum_{P \text{ ext. of } p} T_P(a), \ a \in F.$$

By Lemma 2.2, we then obtain that

$$\prod_{P \text{ ext. of } p} \omega_P(a) = \exp(2\pi i \langle T(a) \rangle_p), \ a \in F.$$

For $\alpha = (a_P)$ in R, let

$$\omega(\alpha) = \prod_{P \in S} \omega_P(a_P).$$

Since $a_P \in \mathfrak{o}_P$ and $\omega_P(a_P) = 1$ for almost all P in S_0, the product is well defined, and $\omega(\alpha)$ is a number in $\mathbf{C}_1$. It is clear that $\omega(\alpha + \beta) = \omega(\alpha)\omega(\beta)$. If $a_P \in \mathfrak{o}_P$ for every P in S_0; and if a_P is near to 0 for every P in S_∞, then $\omega(\alpha) = \prod_{P \in S_\infty} \omega_P(a_P)$ is near to 1. Hence ω is continuous, and ω is a character of R. Let a be in F. Then

$$\begin{aligned}
\omega(a) &= \prod_{P \in S} \omega_P(a) = \prod_p \left(\prod_{P \text{ ext. of } p} \omega_P(a) \right) \prod_{P \in S_\infty} \omega_P(a) \\
&= \prod_p \exp\left(2\pi i \langle T(a) \rangle_p\right) \cdot \exp\left(-2\pi i T(a)\right) \\
&= \exp\left(2\pi i \left(\sum_p \langle T(a) \rangle_p - T(a) \right)\right) \\
&= 1,
\end{aligned}$$

by Lemma 2.2. So, we have the formula

$$\omega(a) = 1, \ a \in F.$$

A character ψ of R such that $\psi(F) = 1$ is called a differential of F, because such characters corresponds to the differentials of an algebraic function field in the analogy between number fields and function fields. The above ω is called the fundamental differential of F.

Let

$$\omega_0(\alpha) = \prod_{P \in S_0} \omega_P(a_P),$$

for $\alpha = (a_P)$ in R. Clearly ω_0 is also a character of R. Since $\omega(a) = 1$ for a in F,

$$\omega_0(a) = \prod_{P \in S_0} \omega_P(a) = \left(\prod_{P \in S_\infty} \omega_P(a) \right)^{-1} = \exp(2\pi i T(a)),$$

namely,

$$\omega_0(a) = \exp(2\pi i T(a)), \ a \in F.$$

We shall next consider a character χ of J, $\chi : J \to \mathbf{C}_1$. Let $\chi_0 : J_0 \to \mathbf{C}_1$ and $\chi_\infty : J_\infty \to \mathbf{C}_1$ be the restrictions of χ on J_0 and J_∞ respectively. For each integral ideal $\mathfrak{f} = \prod_{i=1}^{g} \mathfrak{p}_i^{e_i}$, $e_i \geq 1$ ($\mathfrak{p}_i$ prime ideal of $\mathfrak{o}$), put

$$U_0(\mathfrak{f}) = \prod_{i=1}^{g}(1 + \mathfrak{m}_{P_i}^{e_i}) \times \prod_{\substack{P \neq P_i \\ P \in S_0}} U_P \times 1 \times \cdots \times 1.$$

Then $U_0(\mathfrak{f})$ is an open, compact subgroup of J_0, and

$$U_0(\mathfrak{f}_1 + \mathfrak{f}_2) = U_0(\mathfrak{f}_1)U_0(\mathfrak{f}_2),$$

for any integral ideals $\mathfrak{f}_1$, $\mathfrak{f}_2$. For each $P \in S_0$, the groups $1 + \mathfrak{m}_P^s$, $s = 1, 2, \ldots$, form a base of neighborhoods of 1 in F_P^*. Hence the groups $U_0(\mathfrak{f})$ form a base of neighborhoods of 1 in J_0, when $\mathfrak{f}$ ranges over all integral ideals of F. Let O be a neighborhood of 1 in $\mathbf{C}_1$ which contains no subgroup $\neq 1$. Since χ_0 is continuous, there exists an $\mathfrak{f}_1$ such that $\chi_0(U_0(\mathfrak{f}_1)) \subset O$. By the choice of O, we then have $\chi_0(U_0(\mathfrak{f}_1)) = 1$. Suppose that $\chi_0(U_0(\mathfrak{f}_2)) = 1$ for another integral ideal $\mathfrak{f}_2$. Then $\chi_0(U_0(\mathfrak{f}_1 + \mathfrak{f}_2)) = \chi_0(U_0(\mathfrak{f}_1))\chi_0(U_0(\mathfrak{f}_2)) = 1$. Hence there exists a unique maximal integral ideal $\mathfrak{f}$ such that

$$\chi_0(U_0(\mathfrak{f})) = 1.$$

We call $\mathfrak{f}$ the conductor of χ_0 (and of χ).

Next, we consider χ_∞. A non-zero complex number x can be written in the form

$$x = \frac{x}{|x|} \cdot |x|.$$

This gives us the direct decompositions $\mathbf{R}^* = \{\pm 1\} \times \mathbf{R}^+$, $\mathbf{C}^* = \mathbf{C}_1 \times \mathbf{R}^+$. The characters of $\{\pm 1\}$, $\mathbf{C}_1$, and $\mathbf{R}^+(\cong \mathbf{R})$ can be easily determined. We then see that each character ψ of $\mathbf{R}^*$ is uniquely expressed in the form

$$\psi(x) = \left(\frac{x}{|x|}\right)^m |x|^{is},$$

with $m = 0, 1$ and $s \in \mathbf{R}$, and conversely that any such ψ defines a character of $\mathbf{R}^*$. We also have a similar result for the characters of $\mathbf{C}^*$ with $m = 0, \pm 1, \pm 2, \ldots$.

The isomorphisms $\sigma_k : F_{P_{\infty,k}} \to \mathbf{R}$ or $\mathbf{C}$ induce topological isomorphisms $F_{P_{\infty,k}}^* \to \mathbf{R}^*$ or $\mathbf{C}^*$ for $1 \leq k \leq r$. Let $\alpha = (a_P)$ be in J_∞, and let

$$\sigma(\alpha) = (x_1, \ldots, x_r),\ x_k = \sigma_k(a_{P_{\infty,k}}),\ 1 \leq k \leq r.$$

Then $\sigma : J_\infty \to \mathbf{R}^* \times \cdots \times \mathbf{R}^* \times \mathbf{C}^* \times \cdots \times \mathbf{C}^*$ is also a topological isomorphism. Hence it follows from the above that the character $\chi_\infty : J_\infty \to \mathbf{C}_1$ is of the form

$$\chi_\infty(\alpha) = \prod_{k=1}^{r} \left(\frac{x_k}{|x_k|}\right)^{m'_k} \cdot |x_k|^{is'_k}, \quad \sigma(\alpha) = (x_1, \dots, x_r),$$

where $m'_k = 0, 1$ for $1 \le k \le r_1$, $m'_k = 0, \pm1, \pm2, \dots$ for $r_1 < k \le r$, and $s'_k \in \mathbf{R}$ for $1 \le k \le r$. Let

$$x_{k+r_2} = \overline{x_k} = \overline{\sigma_k}(a_{P_{\infty,k}}), \quad r_1 < k \le r,$$

so that

$$|x_{k+r_2}| = |x_k|, \quad \frac{x_{k+r_2}}{|x_{k+r_2}|} = \left(\frac{x_k}{|x_k|}\right)^{-1}, \quad r_1 < k \le r.$$

Let

$$m_k = m'_k, \qquad s_k = s'_k, \qquad 1 \le k \le r_1,$$

$$\begin{cases} m_k = m'_k, & s_k = \frac{1}{2}s'_k, & r_1 < k \le r, \\ m_{k+r_2} = 0, & s_{k+r_2} = \frac{1}{2}s'_k, & m'_k \ge 0, \end{cases}$$

$$\begin{cases} m_k = 0, & s_k = \frac{1}{2}s'_k, & r_1 < k \le r, \\ m_{k+r_2} = -m'_k, & s_{k+r_2} = \frac{1}{2}s'_k, & m'_k < 0. \end{cases}$$

Then

$$\chi_\infty(\alpha) = \prod_{k=1}^{n} \left(\frac{x_k}{|x_k|}\right)^{m_k} |x_k|^{is_k},$$

where $m_k = 0, 1$ for $1 \le k \le r_1$, $m_k, m_{k+r_2} = 0, 1, 2, \dots$, and $m_k m_{k+r_2} = 0$ for $r_1 < k \le r$, $s_k \in \mathbf{R}$ for $1 \le k \le n$, and $s_k = s_{k+r_2}$ for $r_1 < k \le r$. We call $\{m_1, \dots, m_n; s_1, \dots, s_n\}$ the signature of χ_∞ (and of χ). It determines the character χ_∞, and is uniquely determined for χ_∞ by the above properties.

A character $\chi : J \to \mathbf{C}_1$ such that $\chi(F^* \times T) = 1$ is called a "Grössen" character of Hecke. We shall simply call it a Hecke character of F. Since $J = J_1 \times T$, $J/(F^* \times T) \cong J_1/F^* = \overline{J_1}$, Hecke characters of F are nothing but the characters of $\overline{J_1}$.

Let $\mathfrak{f}$ be an integral ideal of F. Let $\mathfrak{I}(\mathfrak{f})$ denote the set of all ideals $\mathfrak{a}$ in $\mathfrak{I}$ which are prime to $\mathfrak{f}$. Then $\mathfrak{I}(\mathfrak{f})$ is a subgroup of $\mathfrak{I}$. Let $\mathfrak{f} = \prod_{i=1}^{g} \mathfrak{p}_i^{e_i}$, $e_i \ge 1$ ($\mathfrak{p}_i$ prime ideals of $\mathfrak{o}$, distinct), and let $P_i = P_{\mathfrak{p}_i}$, $i = 1, \dots, g$. Put

$$\begin{aligned} J_0(\mathfrak{f}) &= \{\alpha \mid \alpha = (a_P) \in J_0,\ a_{P_i} \in 1 + \mathfrak{m}_{P_i}^{e_i},\ 1 \le i \le g\} \\ &= \{\alpha \mid \alpha = (a_P) \in J_0,\ a_{P_i} \equiv 1 \bmod \mathfrak{m}_{P_i}^{e_i},\ 1 \le i \le g\}. \end{aligned}$$

Then $J_0(f)$ is an open subgroup of J_0 such that

$$J_0(\mathfrak{f}) \cap U = J_0(\mathfrak{f}) \cap U_0 = U_0(\mathfrak{f}).$$

The map $\iota_0 : J_0 \to \mathfrak{I}$ induces a surjective homomorphism $J_0(\mathfrak{f}) \to \mathfrak{I}(\mathfrak{f})$ with kernel $J_0(\mathfrak{f}) \cap U_0 = U_0(\mathfrak{f})$. Hence we have

$$J_0(\mathfrak{f})/U_0(\mathfrak{f}) \cong \mathfrak{I}(\mathfrak{f}).$$

Now, let χ be a character of J with conductor $\mathfrak{f}$. Since $\chi(U_0(\mathfrak{f})) = 1$, there exists a homomorphism $\chi' : \mathfrak{I}(\mathfrak{f}) \to \mathbf{C}_1$ such that

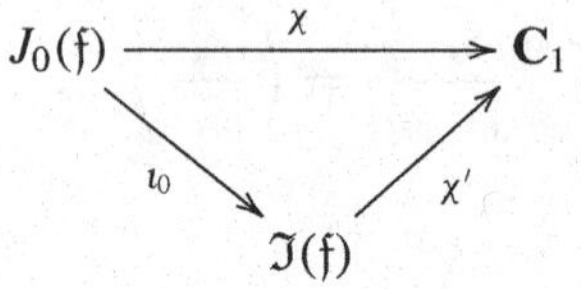

is commutative. We always consider the ideal groups $\mathfrak{I}, \mathfrak{I}(\mathfrak{f})$, etc., in discrete topology. So $\chi' : \mathfrak{I}(\mathfrak{f}) \to \mathbf{C}_1$ is a character of $\mathfrak{I}(\mathfrak{f})$. We call it the ideal character (of $\mathfrak{I}(\mathfrak{f})$) induced by the idèle character χ. If there is no risk of confusion, we shall write $\chi(\mathfrak{a})$ for $\chi'(\mathfrak{a})$, $\mathfrak{a} \in \mathfrak{I}(\mathfrak{f})$. By the definition,

$$\chi(\mathfrak{a}) = \chi(\alpha), \quad \mathfrak{a} \in \mathfrak{I}(\mathfrak{f}),$$

for any idèle α in $J_0(\mathfrak{f})$ such that $\mathfrak{a} = \iota(\alpha)$.

The relation between idèle characters and ideal characters will be studied in detail later on.

2.3 Gaussian Sums

Let $\mathfrak{R}$ be a finite ring with identity 1; for simplicity, we shall assume that $\mathfrak{R}$ is commutative although this is not essential in the following theory. Let $\mathfrak{R}^*$ denote the multiplicative group of all invertible elements in $\mathfrak{R}$. Let λ be a character of the additive group of $\mathfrak{R}$, and ρ a character of the multiplicative group $\mathfrak{R}^*$. Put

$$G(\lambda, \rho) = \sum_{x \in \mathfrak{R}^*} \lambda(x)\rho(x).$$

Such a sum, defined for each pair of characters λ and ρ, is called a Gaussian sum on $\mathfrak{R}$.

Example 2.2 Let $\mathfrak{R} = \mathbf{Z}/p\mathbf{Z} = \{0, 1, \ldots, p-1 \bmod p\}$, $\mathfrak{R}^* = \{1, \ldots, p-1 \bmod p\}$. Let $\lambda(x) = \exp(\frac{2\pi i x}{p})$, $\rho(x) = (\frac{x}{p})$ (Legendre's symbol). Then

$$G(\lambda, \rho) = \sum_{x=1}^{p-1} \left(\frac{x}{p}\right) e^{\frac{2\pi i x}{p}}.$$

For any a in $\mathfrak{R}$, let $\lambda_a(x) = \lambda(ax)$, $x \in \mathfrak{R}$. Then $\lambda_a : \mathfrak{R} \to \mathbf{C}_1$ is a character. Let $\mathfrak{R}_a^* = \{u \mid u \in \mathfrak{R}^*,\ au = a\}$. Then $\mathfrak{R}_a^*$ is a subgroup of $\mathfrak{R}^*$.

Lemma 2.3 (1) $G(\lambda_a, \rho) = \overline{\rho}(a)G(\lambda, \rho)$, *if* $a \in \mathfrak{R}^*$.

(2) $G(\lambda_a, \rho) = 0$, *if* $\rho(\mathfrak{R}_a^*) \neq 1$.

Proof (1) If $a \in \mathfrak{R}^*$, then we have $G(\lambda_a, \rho) = \sum_{x \in \mathfrak{R}^*} \lambda_a(x)\rho(x) = \sum \lambda(ax)\rho(x)$ $= \sum \lambda(x)\rho(a^{-1}x) = \rho(a)^{-1} \sum \lambda(x)\rho(x) = \overline{\rho}(a)G(\lambda, \rho)$.

(2) Let $\mathfrak{R}^* = \bigcup_{i=1}^h \mathfrak{R}_a^* x_i$, $h = [\mathfrak{R}^* : \mathfrak{R}_a^*]$ so that $x = ux_i$, $u \in \mathfrak{R}_a^*$, $1 \leq i \leq h$, for any x in $\mathfrak{R}^*$. Then $G(\lambda_a, \rho) = \sum_{i=1}^h \sum_{u \in \mathfrak{R}_a^*} \lambda(aux_i)\rho(ux_i) = \sum_i \lambda(ax_i)\rho(x_i)$ $\sum_u \rho(u)$. However, since $\rho(\mathfrak{R}_a^*) \neq 1$, $\sum_u \rho(u) = 0$. Hence $G(\lambda_a, \rho) = 0$.

Lemma 2.4 *Suppose that*

(i) $\lambda_a(\mathfrak{R}) \neq 1$ *for any* a *in* $\mathfrak{R}$, $a \neq 0$, *and*
(ii) $\rho(\mathfrak{R}_a^*) \neq 1$ *for any* a *in* $\mathfrak{R}$, $a \notin \mathfrak{R}^*$.

Then

$$|G(\lambda, \rho)|^2 = \#(\mathfrak{R}).$$

In particular,

$$G(\lambda, \rho) = 0.$$

Proof $|G(\lambda, \rho)|^2 = \sum_{x \in \mathfrak{R}^*} \lambda(x)\rho(x) \sum_{y \in \mathfrak{R}^*} \overline{\lambda}(y)\overline{\rho}(y) = \sum_{x,y} \lambda(x-y)\rho(\frac{x}{y})$. Let $z = \frac{x}{y}$. Then the sum is equal to

$$\sum_{y,z \in \mathfrak{R}^*} \lambda((z-1)y)\rho(z) = \sum_{z \in \mathfrak{R}^*} \left(\sum_{y \in \mathfrak{R}} - \sum_{\substack{y \in \mathfrak{R} \\ y \notin \mathfrak{R}^*}} \right).$$

However,

$$\sum_{z \in \mathfrak{R}^*} \sum_{y \in \mathfrak{R}} = \sum_{z} \rho(z) \sum_{y \in \mathfrak{R}} \lambda_{z-1}(y) = \rho(1)\#(\mathfrak{R}) = \#(\mathfrak{R}),$$

because $\lambda_{z-1} \not\equiv 1$ for $z \neq 1$. On the other hand,

$$\sum_{z \in \mathfrak{R}^*} \sum_{y \notin \mathfrak{R}^*} = \sum_{y \notin \mathfrak{R}^*} \lambda(-y) \sum_{z \in \mathfrak{R}^*} \lambda(zy)\rho(z) = \sum_{y \notin \mathfrak{R}^*} \lambda(-y)G(\lambda_y, \rho).$$

By the assumption (ii) and by Lemma 2.3 (1), we have $G(\lambda_y, \rho) = 0$ for every y in $\mathfrak{R}$, $y \notin \mathfrak{R}^*$. Hence the above sum is 0, and the lemma is proved.

Example 2.3 For $G(\lambda, \rho) = \sum_{x=1}^{p-1} (\frac{x}{p}) e^{\frac{2\pi i x}{p}}$, the assumptions (i), (ii) are satisfied for λ, ρ. Hence

$$|G(\lambda, \rho)|^2 = \#(\mathfrak{R}) = p.$$

Let $\mathfrak{a}$ be an ideal of F, $\mathfrak{a} \neq \{0\}$, $\mathfrak{a} = \prod \mathfrak{p}^{\nu_\mathfrak{p}(\mathfrak{a})}$. Put

$$\begin{aligned} R_\mathfrak{a} &= \prod_\mathfrak{p} \mathfrak{m}_\mathfrak{p}^{\nu_\mathfrak{p}(\mathfrak{a})} \times \prod_{P \in S_\infty} F_P \\ &= \{\alpha \mid \alpha = (a_P),\ a_P \in \mathfrak{m}_P^{\nu_P(\mathfrak{a})},\ \text{ for all } P \in S_0\} \\ &= \{\alpha \mid \alpha = (a_P),\ \nu_P(a_P) \geq \nu_P(\mathfrak{a}),\ \text{ for all } P \in S_0\}. \end{aligned}$$

For $\mathfrak{a} = \mathfrak{o}$, $R_\mathfrak{a} = \prod \mathfrak{o}_\mathfrak{p} \times \prod F_P = R'$ (in Sect. 2.1). In general, $R_\mathfrak{a}$ is an open R'-submodule of R, and $F \cap R_\mathfrak{a} = \mathfrak{a}$. If $\mathfrak{a} \subset \mathfrak{a}'$, then $R_\mathfrak{a} \subset R_{\mathfrak{a}'}$.

Let $\mathfrak{a}$ be integral. Then $R_\mathfrak{a} \subset R' = R_\mathfrak{o}$, and $R_\mathfrak{a}$ is an ideal of R'. Let

$$\mathfrak{R}_\mathfrak{a} = R'/R_\mathfrak{a}, \quad \mathfrak{a} \subset \mathfrak{o}.$$

Proposition 2.3 (1) *There is a natural isomorphism*

$$\mathfrak{R}_\mathfrak{a} \cong \mathfrak{o}/\mathfrak{a},$$

so that $\mathfrak{R}_\mathfrak{a}$ is a finite ring with $N(\mathfrak{a})$ elements.

(2) *Let $\alpha = (a_P)$ be an element of R', and let $\overline{\alpha}$ be the residue class of α in $\mathfrak{R}_\mathfrak{a} = R'/R_\mathfrak{a}$. Then $\overline{\alpha}$ is in $\mathfrak{R}_\mathfrak{a}^*$ if and only if $\nu_P(a_P) = 0$ for every $P \in S_0$ such that $\nu_P(\mathfrak{a}) > 0$.*

Proof (1) $\mathfrak{a} = \mathfrak{o} \cap \mathfrak{a} = \mathfrak{o} \cap F \cap R_\mathfrak{a} = \mathfrak{o} \cap R_\mathfrak{a}$. Let $\alpha = (a_P)$ be in R'. By the strong approximation theorem, there exists a in F such that $a_P \equiv a \bmod \mathfrak{m}_P^{\nu_P(\mathfrak{a})}$ for every $P \in S_0$. Let $\beta = \alpha - a$. Then β is in $R_\mathfrak{a}$. Hence $a = \alpha - \beta$ is in R', and consequently in $\mathfrak{o} = F \cap R'$. Thus $\alpha = a + \beta$ with $a \in \mathfrak{o}$, $\beta \in R_\mathfrak{a}$. Therefore $R' = \mathfrak{o} + R_\mathfrak{a}$, and together with $\mathfrak{a} = \mathfrak{o} \cap R_\mathfrak{a}$, we obtain $\mathfrak{R}_\mathfrak{a} = R'/R_\mathfrak{a} \cong \mathfrak{o}/\mathfrak{a}$.

(2) Suppose that $\overline{\alpha} \in \mathfrak{R}_\mathfrak{a}^*$. Then $\overline{\alpha}\overline{\beta} = 1$, namely, $\alpha\beta \equiv 1 \bmod R_\mathfrak{a}$ for some $\beta = (b_P)$ in R'. It follows that $\nu_P(a_P b_P - 1) \geq \nu_P(\mathfrak{a})$ for every $P \in S_0$. Suppose that $\nu_P(\mathfrak{a}) > 0$. Then $\nu_P(a_P b_P - 1) > 0$ so that $\nu_P(a_P b_P) = 0$. Since $\nu_P(a_P), \nu_P(b_P) \geq 0$, we obtain $\nu_P(a_P)(= \nu_P(b_P)) = 0$.

Suppose conversely that $\nu_P(a_P) = 0$ for every $P \in S_0$ such that $\nu_P(\mathfrak{a}) > 0$. Let $\beta = (b_P)$ where $b_P = a_P^{-1}$ if $P \in S_0$ and $\nu_P(\mathfrak{a}) > 0$, and $b_P = 0$ otherwise. Then β is in R', and $\nu_P(a_P b_P - 1) \geq \nu_P(\mathfrak{a})$ for any $P \in S_0$. Hence $\alpha\beta \equiv 1 \bmod R_\mathfrak{a}$, i.e. $\overline{\alpha}\overline{\beta} = 1$, and thus $\overline{\alpha}$ is in $\mathfrak{R}_\mathfrak{a}^*$.

Problem 2.3 Prove that

$$\mathfrak{R}_\mathfrak{a}^* \cong U_0/U_0(\mathfrak{a}), \quad \mathfrak{a} \subset \mathfrak{o}.$$

Let $\mathfrak{a}$ be an ideal of F, $\mathfrak{a} \neq \{0\}$. An element α in J is called an idèle associated with $\mathfrak{a}$ if

$$\iota(\alpha_0) = \mathfrak{a}, \quad \alpha_\infty = \tau_{N(\mathfrak{a})},$$

for $\alpha = \alpha_0\alpha_\infty$, $\alpha_0 \in J_0$, $\alpha_\infty \in J_\infty$. If $\alpha = (a_P)$, the condition is equivalent to that

$$\nu_P(a_P) = \nu_P(\mathfrak{a}) \text{ for } P \in S_0, \quad a_{P_{\infty,i}} = \sigma_i^{-1}(N(\mathfrak{a})^{\frac{1}{n}}) \text{ for } P_{\infty,i} \in S_\infty.$$

Since $\iota_0 : J_0 \to \mathfrak{J}$ is surjective, such an α always exists, and if α' is also associated with $\mathfrak{a}$, then $\alpha' = \alpha\xi$, $\xi \in U_0$, and conversely

Problem 2.4 Prove that α is associated with $\mathfrak{a}$ if and only if

$$\iota(\alpha) = \mathfrak{a}, \quad \alpha \in (J_0 \times T) \cap J_1.$$

Now, let $\mathfrak{d}$ be as before the different of F, and let $d = N(\mathfrak{d}) = |\Delta|$, where Δ is the discriminant of F. Let δ be an idèle associated with $\mathfrak{d}$, namely, $\delta = (d_P)$, $\nu_P(d_P) = \nu_P(\mathfrak{d})$, $P \in S_0$, $d_{P_{\infty,i}} = \sigma_i^{-1}(d^{\frac{1}{n}})$, $P_{\infty,i} \in S_\infty$. Let $\mathfrak{d}_P$ be the different of F_P, $P \in S_0$. Then $\mathfrak{d}_P = \mathfrak{m}_P^{\nu_P(\mathfrak{d})}$ (cf. Sect. 1.2). Hence

$$\mathfrak{d}_P = d_P\mathfrak{o}_P, \quad \widetilde{\mathfrak{o}_P} = \mathfrak{d}_P^{-1} = \frac{1}{d_P}\mathfrak{o}_P, \quad P \in S_0.$$

Let $\chi : J \to \mathbf{C}_1$ be a character of J, and let $\mathfrak{f}$ be the conductor of χ. Let φ be an idèle associated with $\mathfrak{f}$, namely, $\varphi = (f_P)$, $\nu_P(f_P) = \nu_P(\mathfrak{f})$, $P \in S_0$, $f_{P_{\infty,i}} = \sigma_i^{-1}(f^{\frac{1}{n}})$, $P_{\infty,i} \in S_\infty$ where $f = N(\mathfrak{f})$. Let

$$\mathfrak{R} = \mathfrak{R}_\mathfrak{f} = R'/R_\mathfrak{f} \cong \mathfrak{o}/\mathfrak{f}.$$

Then $\mathfrak{R}$ is a finite ring with f elements. We shall fix χ, φ, and δ, and define a Gaussian sum on the ring $\mathfrak{R}$.

I. The Additive Character $\lambda : \mathfrak{R} \to \mathbf{C}_1$

Let $\omega_0 : R \to \mathbf{C}_1$ be the character defined in Sect. 2.2. Define $\omega_0' : R' \to \mathbf{C}_1$ by

$$\omega_0'(\xi) = \omega_0\left(\frac{\xi}{\delta\varphi}\right), \quad \xi \in R'.$$

Since δ, φ are in J, $\frac{\xi}{\delta\varphi}$ is in R. Hence ω_0' is well defined, and is obviously a character of R'.

Let $\alpha = (a_P)$ be an element of R'. We shall prove that $\omega_0'(\alpha\xi) = 1$ for every $\xi = (x_P)$ in R' $\Leftrightarrow$ α is in $R_\mathfrak{f}$. Suppose that $\omega_0'(\alpha\xi) = 1$ for any $\xi = (x_P)$ in R'. Then

$$\omega_0'(\alpha\xi) = \omega_0\left(\frac{\alpha\xi}{\delta\varphi}\right) = \prod_{P \in S_0} \omega_P\left(\frac{a_P x_P}{d_P f_P}\right) = 1.$$

Let P be fixed, $P \in S_0$, and let $\xi = (x_p, 0, \ldots, 0)$ with arbitrary x_P in $\mathfrak{o}_P$. Then ξ is in R', and it follows from the above that

$$\omega_P\left(\frac{a_P x_P}{d_P f_P}\right) = \exp\left(2\pi i \left\langle T_p\left(\frac{a_P x_P}{d_P f_P}\right)\right\rangle_p\right) = 1,$$

namely, that $T_p(\frac{a_P x_P}{d_P f_P})$ is contained in $\mathbf{Z}_p$ (p = restriction of P on $\mathbf{Q}$). Since this holds for any x_P in $\mathfrak{o}_P$, we see that $\frac{a_P}{d_P f_P} \in \widetilde{\mathfrak{o}_P} = \frac{1}{d_P}\mathfrak{o}_P$, namely, that $a_P \in f_P \mathfrak{o}_P$, $\nu_P(a_P) \geq \nu_P(f_P) = \nu_P(\mathfrak{f})$. As this is true for arbitrary P in $S_0, \alpha = (a_P)$ is contained in $R_\mathfrak{f}$. Conversely, if $\alpha = (a_P)$ is in $R_\mathfrak{f}$, then $\nu_P(a_P) \geq \nu_P(\mathfrak{f})$, $P \in S_0$, and the above computation shows that $\omega_P(\frac{a_P x_P}{d_P f_P}) = 1$ for any x_P in $\mathfrak{o}_P$. Hence $\omega_0'(\alpha\xi) = 1$ for any $\xi = (x_P)$ in R'.

In particular, it follows from the above that $\omega_0'(\alpha) = 1$ if α is in $R_\mathfrak{f}$. Hence $\omega_0'(\alpha)$ depends only upon the residue class $\overline{\alpha}$ of α mod $R_\mathfrak{f}$, and we obtain a character $\lambda : \mathfrak{R} = R'/R_\mathfrak{f} \to \mathbf{C}_1$ by putting

$$\lambda(\overline{\alpha}) = \omega_0'(\alpha), \quad \overline{\alpha} \in \mathfrak{R} \text{ (i.e. } \alpha \in R').$$

The above result then shows that $\lambda_{\overline{\alpha}}(\overline{\xi}) = \lambda(\overline{\alpha}\overline{\xi}) = \omega_0'(\alpha\xi) = 1$ for any $\overline{\xi}$ in $\mathfrak{R}$ (for any ξ in R') $\Leftrightarrow \overline{\alpha} = 0$. Thus

$$\lambda_{\overline{\alpha}}(\mathfrak{R}) \neq 1, \quad \text{for any } \overline{\alpha} \text{ in } \mathfrak{R},\ \overline{\alpha} \neq 0.$$

II. The Multiplicative Character $\rho : \mathfrak{R}^* \to \mathbf{C}_1$

For any $\alpha = (a_P)$ in R, let $\alpha_\mathfrak{f} = (a'_P)$ be defined by

$$a'_P = \begin{cases} a_P, & \text{if } P \in S_0 \text{ and } \nu_P(\mathfrak{f}) > 0, \\ 1, & \text{otherwise.} \end{cases}$$

Then $\alpha \mapsto \alpha_\mathfrak{f}$ is a continuous map $R \to R$ such that $(\alpha\beta)_\mathfrak{f} = \alpha_\mathfrak{f}\beta_\mathfrak{f}$. The restriction on J defines a continuous homomorphism $J \to J$. Note that if $\mathfrak{f} = 0$, then $\alpha_\mathfrak{f} = 1$ for any α in R.

Define $\chi_\mathfrak{f} : R \to \mathbf{C}$ by

$$\chi_\mathfrak{f}(\alpha) = \begin{cases} \chi(\alpha_\mathfrak{f}), & \text{if } \alpha_\mathfrak{f} \in U_0 \text{ (or } \alpha_\mathfrak{f} \in U), \\ 0, & \text{otherwise.} \end{cases}$$

For example, $\chi_\mathfrak{f}(0) = 1$ or 0 according as $\mathfrak{f} = \mathfrak{o}$ or $\mathfrak{f} \neq \mathfrak{o}$. Also, if α is in U_0, then $\alpha_\mathfrak{f} \in U_0, \alpha_\mathfrak{f}/\alpha \in U_0(\mathfrak{f})$ so that

$$\chi_\mathfrak{f}(\alpha) = \chi(\alpha_\mathfrak{f}) = \chi(\alpha(\alpha_\mathfrak{f}/\alpha)) = \chi(\alpha).$$

Let $\alpha = (a_P)$ be in R', and let $\overline{\alpha}$ be the residue class of α mod $R_\mathfrak{f}$, $\overline{\alpha} \in \mathfrak{R}$. We shall next prove that $\chi_\mathfrak{f}(\alpha)$ depends only upon $\overline{\alpha}$. It follows from Proposition 2.3 that

$$\chi_{\mathfrak{f}}(\alpha) = \chi(\alpha_{\mathfrak{f}}) \text{ or } \chi_{\mathfrak{f}}(\alpha) = 0$$

according as $\overline{\alpha} \in \mathfrak{R}^*$ or $\overline{\alpha} \notin \mathfrak{R}^*$. Hence it is sufficient to show that $\chi(\alpha_{\mathfrak{f}}) = \chi(\beta_{\mathfrak{f}})$ if $\overline{\alpha} = \overline{\beta}$, $\beta = (b_P) \in R'$. Let $\nu_P(\mathfrak{f}) > 0$ so that $\nu_P(a_P) = \nu_P(b_P) = 0$. Since $\alpha \equiv \beta \bmod R_{\mathfrak{f}}$, $\nu'_P(a_P - b_P) \geq \nu_P(\mathfrak{f})$. Hence $\nu_P(\frac{a_P}{b_P} - 1) = \nu_P(a_P - b_P) - \nu_P(b_P) \geq \nu_P(\mathfrak{f})$ for any $P \in S_0$ with $\nu_P(\mathfrak{f}) > 0$. Therefore $\frac{\alpha_{\mathfrak{f}}}{\beta_{\mathfrak{f}}}$ is contained in $U_0(\mathfrak{f})$, and $\chi(\frac{\alpha_{\mathfrak{f}}}{\beta_{\mathfrak{f}}}) = 1$, namely $\chi(\alpha_{\mathfrak{f}}) = \chi(\beta_{\mathfrak{f}})$.

We now define $\rho : \mathfrak{R}^* \to \mathbf{C}_1$ by

$$\rho(\overline{\alpha}) = \overline{\chi}_{\mathfrak{f}}(\alpha) = \chi_{\mathfrak{f}}(\alpha)^{-1} = \chi(\alpha_{\mathfrak{f}})^{-1}, \ \overline{\alpha} \in \mathfrak{R}^*,$$

where α is any element of R' in the residue class $\overline{\alpha}$. By the above remarks, ρ is well-defined, and gives a character of $\mathfrak{R}^*$.

Let $\overline{\alpha} \in \mathfrak{R}, \overline{\alpha} \notin \mathfrak{R}^*$. Let $\alpha = (a_P)$ be in the residue class $\overline{\alpha}$. Then there exists a spot $P_0 \in S_0$ such that $\nu_{P_0}(a_{P_0}) > 0$, $\nu_{P_0}(\mathfrak{f}) > 0$ (Proposition 2.3). Let $P_0 \leftrightarrow \mathfrak{p}_0$, $\mathfrak{f}' = \mathfrak{f}/\mathfrak{p}_0$. Since $\nu_{P_0}(\mathfrak{f}') = \nu_{P_0}(\mathfrak{f}) - 1 \geq 0$, $\mathfrak{f}'$ is an integral ideal of F, $\mathfrak{f} \subset \mathfrak{f}'$, $\mathfrak{f} \neq \mathfrak{f}'$. Hence $\chi(U_0(\mathfrak{f})) \neq 1$ by the definition of the conductor $\mathfrak{f}$. Let $\xi = (x_P) \in U_0(\mathfrak{f}')$, $\chi(\xi) \neq 1$. Since ξ is in U_0, $\chi_{\mathfrak{f}}(\xi) = \chi(\xi) \neq 1$. Hence $\overline{\xi}$ is in $\mathfrak{R}^*$, and $\rho(\overline{\xi}) = \overline{\chi}_{\mathfrak{f}}(\xi) \neq 1$. On the other hand, $\nu_P(a_P x_P - a_P) = \nu_P(a_P) + \nu(x_P - 1) \geq \nu_P(a_P) + \nu_P(\mathfrak{f}')$, $P \in S_0$. If $P \neq P_0$, then $\nu_P(\mathfrak{f}') = \nu_P(\mathfrak{f})$ and $\nu_P(a_P) \geq 0$ $(\alpha \in R')$ so that $\nu_P(a_P x_P - a_P) \geq \nu_P(\mathfrak{f})$. If $P = P_0$, then $\nu_P(\mathfrak{f}') = \nu_P(\mathfrak{f}) - 1$, but $\nu_P(a_P) \geq 1$ so that again $\nu_P(a_P x_P - a_P) \geq \nu_P(\mathfrak{f})$. Hence $\alpha\xi - \alpha$ is in $R_{\mathfrak{f}}$, namely, $\overline{\alpha}\overline{\xi} = \overline{\alpha}$. Therefore $\overline{\xi}$ belongs to the subgroup $\mathfrak{R}^*_{\overline{\alpha}}$ of $\mathfrak{R}^*$, and it follows from the above that

$$\rho(\mathfrak{R}^*_{\overline{\alpha}}) \neq 1, \quad \overline{\alpha} \notin \mathfrak{R}^*.$$

Thus the condition (ii) of Lemma 2.4 is satisfied for the character ρ.

We now put

$$\begin{aligned} C(\chi; \delta, \varphi) &= G(\lambda, \rho) \\ &= \sum_{\overline{\xi} \in \mathfrak{R}^*} \lambda(\overline{\xi})\rho(\overline{\xi}), \end{aligned}$$

with the characters $\lambda : \mathfrak{R} \to \mathbf{C}_1$ and $\rho : \mathfrak{R}^* \to \mathbf{C}_1$ defined in the above. Note that if $\mathfrak{f} = \mathfrak{o}$, then $R_{\mathfrak{f}} = R_{\mathfrak{o}} = R'$ so that $\mathfrak{R} = \{0\}$, $\mathfrak{R}^* = \{1\} = \{0\} = \mathfrak{R}$, and $C(\chi; \delta, \varphi) = 1$.

Proposition 2.4

$$|C(\chi; \delta, \varphi)| = \sqrt{f}, \quad f = N(\mathfrak{f}),$$
$$\overline{C(\chi; \delta, \varphi)} = \chi_{\mathfrak{f}}(-1)C(\overline{\chi}; \delta, \varphi).$$

Proof The first equality follows immediately from Lemma 2.4. As to the second,

$$\overline{C(\chi;\delta,\varphi)} = \sum \overline{\lambda}(\overline{\xi})\overline{\rho}(\overline{\xi}) = \sum \lambda(-\overline{\xi})\overline{\rho}(\overline{\xi})$$
$$= \sum \lambda(\overline{\xi})\overline{\rho}(-\overline{\xi}) = \overline{\rho}(-1)\sum \lambda(\overline{\xi})\overline{\rho}(\overline{\xi}).$$

Here $\overline{\rho}(-1) = \chi_{\mathfrak{f}}(-1), \overline{\rho}(\overline{\xi}) = \overline{\chi}_{\mathfrak{f}}(\xi)$. Hence $\sum \lambda(\overline{\xi})\overline{\rho}(\overline{\xi}) = G(\lambda, \overline{\rho}) = C(\overline{\chi};\delta,\varphi)$, and the equality is proved.

Let $\mathfrak{a}$ and $\mathfrak{b}$ be ideals of F, $\mathfrak{a}$, $\mathfrak{b} \neq \{0\}$, $\mathfrak{b} \subset \mathfrak{a}$, and let $\xi \in J$. Then

1. $R_{\mathfrak{a}} = \mathfrak{a} + R_{\mathfrak{b}}$, $\mathfrak{b} = \mathfrak{a} \cap R_{\mathfrak{b}}$ so that $\mathfrak{a}/\mathfrak{b} \cong R_{\mathfrak{a}}/R_{\mathfrak{b}}$ under $x + \mathfrak{b} \leftrightarrow x + R_{\mathfrak{b}}$, $x \in \mathfrak{a}$,
2. $\xi R_{\mathfrak{a}} = R_{\iota(\xi)\mathfrak{a}}$ so that $R_{\mathfrak{a}} \cong R_{\iota(\xi)\mathfrak{a}}$ under $\alpha \leftrightarrow \xi\alpha$, $\alpha \in R_{\mathfrak{a}}$.

Problem 2.5 Prove the above 1, 2.

We now prove the following.

Proposition 2.5 *Let* $\alpha \in J$, $\mathfrak{a} = \iota(\alpha)$, $a \in \mathfrak{a}\mathfrak{d}^{-1}\mathfrak{f}^{-1}$ $(b \in F)$. *Then*

$$\sum_{x} \omega_0(xa)\overline{\chi}_{\mathfrak{f}}(x\alpha) = \chi_{\mathfrak{f}}(\alpha^{-1}\delta\varphi a)C(\chi;\delta,\varphi),$$

where x ranges over a set of representatives of $\mathfrak{a}^{-1}$ modulo $\mathfrak{a}^{-1}\mathfrak{f}$.

Proof When x moves as stated, then x also ranges over a set of representatives of $R_{\mathfrak{a}^{-1}}$ mod $R_{\mathfrak{a}^{-1}\mathfrak{f}}$ (see 1. in the above). Hence $\xi = x\alpha$ ranges over a set of representatives of $\alpha R_{\mathfrak{a}^{-1}} = R_{\iota(\alpha)\mathfrak{a}^{-1}} = R_{\mathfrak{o}} = R'$ modulo $\alpha R_{\mathfrak{a}^{-a}\mathfrak{f}} = R_{\iota(\alpha)\mathfrak{a}^{-1}\mathfrak{f}} = R_{\mathfrak{f}}$. Hence the left hand side of the equality in the proposition is equal to

$$\sum_{\xi \in \mathfrak{R}} \omega\left(\frac{a\xi}{\alpha}\right)\overline{\chi}_{\mathfrak{f}}(\xi) = \sum_{\xi \in \mathfrak{R}} \omega_0\left(\frac{\alpha'\xi}{\delta\varphi}\right)\overline{\chi}_{\mathfrak{f}}(\xi),$$

where $\alpha' = \alpha^{-1}\delta\varphi a$. Let $\alpha = (a_P)$, $\alpha' = (a'_P)$. Then $a'_P = a_P^{-1}d_P f_P a$ so that $\nu_P(a'_P) = -\nu_P(a_P) + \nu_P(d_P) + \nu_P(f_P) + \nu_P(a) = -\nu_P(\mathfrak{a}) + \nu_P(\mathfrak{d}) + \nu_P(\mathfrak{f}) + \nu_P(a) = \nu_P(\mathfrak{a}^{-1}\mathfrak{d}\mathfrak{f}a) \geq 0$ because $a \in \mathfrak{a}\mathfrak{d}^{-1}\mathfrak{f}^{-1}$, $a\mathfrak{a}^{-1}\mathfrak{d}\mathfrak{f} \subset \mathfrak{o}$. Hence α' is in R'. As $\overline{\chi}_{\mathfrak{f}}(\xi) = 0$ if $\overline{\xi} \notin \mathfrak{R}^*$, we see that the above sum is equal to

$$\sum_{\overline{\xi} \in \mathfrak{R}^*} \lambda(\overline{\alpha}'\overline{\xi})\rho(\overline{\xi}) = \sum_{\overline{\xi}} \lambda_{\overline{\alpha}'}(\overline{\xi})\rho(\overline{\xi})$$
$$= G(\lambda_{\overline{\alpha}'}, \rho).$$

Suppose that $\overline{\alpha}' \in \mathfrak{R}^*$. Then it follows from Lemma 2.3 that

$$G(\lambda_{\overline{\alpha}'}, \rho) = \overline{\rho}(\overline{\alpha}')G(\lambda, \rho) = \chi_{\mathfrak{f}}(\alpha')C(\chi;\delta,\varphi) = \chi_{\mathfrak{f}}(\alpha^{-1}\delta\varphi a)C(\chi;\delta,\varphi).$$

Suppose that $\overline{\alpha}' \notin \mathfrak{R}^*$. Then $\rho(\mathfrak{R}^*_{\overline{\alpha}'}, \rho) \neq 1$. Hence it follows again from Lemma 2.3 that $G(\lambda'_{\overline{\alpha}}, \rho) = 0$. However, in this case, $\chi_{\mathfrak{f}}(\alpha') = 0$. Hence the proposition is proved.

2.4 Haar Measures

We shall next state without proof some fundamental results on Haar measures of locally compact groups. For simplicity, we shall always assume that our locally compact groups are abelian and separable (i.e. have a countable basis for open sets).

Problem 2.6 Prove that R and J are separable.

Let G be such a locally compact, separable, abelian group. A Borel (=Baire) measure μ on G is a measure defined on the family of Borel (=Baire) subsets of G such that $\mu(C) < +\infty$ whenever C is compact. A Borel measure μ on G is called a Haar measure if $\mu \not\equiv 0$ and if $\mu(aA) = \mu(A)$ for any a in G and for any Borel subset A of G. The fundamental theorem states that there exists such a Haar measure on G, and that it is unique up to a constant factor.

1. Let μ be a Haar measure on G. Let $f(x) \in L^1(G, \mu)$. Then $f(x^{-1}), f(a^{-1}x) \in L^1(G, \mu)$, $a \in G$, and

$$\int_G f(x)d\mu(x) = \int_G f(x^{-1})d\mu(x) = \int_G f(a^{-1}x)d\mu(x),$$

namely,

$$d\mu(x) = d\mu(x^{-1}) = d\mu(ax), \quad a \in G.$$

2. G is compact if and only if $\mu(G) < +\infty$ for a Haar measure μ on G. In such a case, for any given $c > 0$, there exists a unique Haar measure μ on G such that $\mu(G) = c$.

3. G is discrete if and only if each point of G has a positive measure for a Haar measure of G. In such a case, there exists a unique Haar measure μ on G such that $\mu(x) = 1$ for every x in G. We shall call such μ the point measure on G.

4. Let $\sigma : G \to G'$ be a topological isomorphism. Let μ be a Haar measure on G. Define μ' by $\mu'(A') = \mu(\sigma^{-1}(A'))$ for any Borel subset A' of G'. Then μ' is a Haar measure of G'. We denote it by μ^σ. We also write $\mu \underset{\sigma}{\approx} \mu'$ when μ and μ' are related as mentioned in the above.

5. Let μ_1, μ_2 be Haar measures on G_1 and G_2 respectively. Then the product measure $\mu_1 \times \mu_2$ is a Haar measure on $G_1 \times G_2$.

Let H be a closed subgroup of G. Then both H and G/H are again locally compact, separable, abelian groups. Let μ_G, μ_H, and $\mu_{G/H}$ be Haar measures of the respective groups. Let f be a continuous function with compact support on G. Then $g(x) = \int_H f(xu)d\mu_H(u)$ is defined for any x in G, and $g(x)$ depends only upon the coset $x' = xH$. Hence we put

$$g'(x') = \int_H f(xu)d\mu_H(x), \quad x' = xH \in G/H.$$

The function g' is then continuous, and has a compact support in G/H. Hence $\int_{G/H} g'(x')d\mu_{G/H}(x')$ is defined.

Now, we say that the measures μ_G, μ_H, and $\mu_{G/H}$ are matched if

$$\int_G f(x)d\mu_G(x) = \int_{G/H} g'(x')d\mu_{G/H}(x')$$
$$= \int_{G/H} \left(\int_H f(xu)d\mu_H(u) \right) d\mu_{G/H}(x'), \qquad (2.1)$$

for any function f which is continuous and has a compact support on G.

Given Haar measures on any two of the groups G, H, and G/H, there exists a unique Haar measure on the third group such that the three measures thus obtained are matched in the above sense. If μ_G, μ_H, and $\mu_{G/H}$ are matched, we write

$$\mu_G = \mu_{G/H} * \mu_H, \quad \mu_{G/H} = \mu_G/\mu_H.$$

6. Let $H \subset G, \mu_G = \mu_{G/H} * \mu_H$. Let f be a non-negative measurable function on G. Then $g'(x') = \int_H f(xu)d\mu_H(u)$ is also non-negative and measurable on G/H, and (2.1) holds.

7. In the same situation, let f be a function in $L^1(G, \mu_G)$. Then

$$g'(x') = \int_H f(xu)d\mu_H(x)$$

is defined for almost all x' in G/H; g' is a function in $L^1(G/H, \mu_{G/H})$, and (2.1) again holds.

Let H be a closed subgroup of G; and U a closed subgroup of H. Let $\mu_{G/H}$ and $\mu_{H/U}$ be Haar measures of the respective groups. Let $G' = G/U$, $H' = H/U$. Then G/H is naturally identified with G'/H' so that $\mu_{G/H} = \mu_{G'/H'}$. As $\mu_{H/U} = \mu_{H'}$, $\mu_{G/H} * \mu_{H/U} = \mu_{G'/H'} * \mu_{H'}$ is defined, and it is a Haar measure on G/U.

8. Let $V \subset U \subset H \subset G$ be a sequence of closed subgroups, and let $\mu_{G/H}, \mu_{H/U}$, $\mu_{U/V}$ be given Haar measures on the respective groups. Then Associative law: $\mu_{G/H} * (\mu_{H/U} * \mu_{U/V}) = (\mu_{G/H} * \mu_{H/U}) * \mu_{U/V}$.

9. Let $U \subset H \subset G$, and let μ_G, μ_H, and μ_U be given. Then

$$(\mu_G/\mu_U)/(\mu_H/\mu_U) = \mu_G/\mu_H.$$

This follows from 7 by putting $V = 1, \mu_{G/H} = \mu_G/\mu_H, \mu_{H/U} = \mu_H/\mu_U, \mu_{U/V} = \mu_U$.

10. Let $H_1 \subset G_1$, $H_2 \subset G_2$, and let $\mu_{G_1}, \mu_{G_2}, \mu_{H_1}, \mu_{H_2}$ be given. Then

$$(\mu_{G_1} \times \mu_{G_2})/(\mu_{H_1} \times \mu_{H_2}) \approx (\mu_{G_1}/\mu_{H_1}) \times (\mu_{G_2}/\mu_{H_2})$$

under the natural isomorphism $(G_1 \times G_2)/(H_1 \times H_2) \cong (G_1/H_1) \times (G_2/H_2)$.

11. Let μ_{G_1} and μ_{G_2} be given. Then

$$(\mu_{G_1} \times \mu_{G_2})/\mu_{G_1} \approx \mu_{G_2}, \quad (\mu_{G_1} \times \mu_{G_2})/\mu_{G_2} \approx \mu_{G_1},$$

under the natural isomorphisms $(G_1 \times G_2)/G_1 \cong G_2$, $(G_1 \times G_2)/G_2 \cong G_1$.

We also note that given $\mu_{G_1 \times G_2}$ and $\mu_{G_1}(\mu_{G_2})$, there exists a unique $\mu_{G_2}(\mu_{G_1})$ such that $\mu_{G_1 \times G_2} = \mu_{G_1} \times \mu_{G_2}$.

Example 2.4 1. $G = \mathbf{R}^n$, $x = (x_1, \ldots, x_n) \in \mathbf{R}^n$, $n \geq 1$.

$$d\mu(x) = dx_1 \cdots dx_n.$$

2. $G = \mathbf{R}^*$ or $\mathbf{R}^+$, $x \in \mathbf{R}^*$ or $\mathbf{R}^+$

$$d\mu(x) = \frac{dx}{|x|}.$$

3. $G = \mathbf{C}^*$, $z \in \mathbf{C}^*$, $z = x + iy = re^{i\theta}$.

$$d\mu(z) = \frac{2dxdy}{|z|^2} = \frac{2drd\theta}{r}.$$

The Haar measures of $\mathbf{R}^n$, $\mathbf{R}^*$, $\mathbf{R}^+$, and $\mathbf{C}^*$ as normalized in the above, will be called the standard measures on these groups.

We shall next fix Haar measures of various groups defined by J and its subgroups.

Let μ_{U_0} be the Haar measure of the compact group U_0 such that $\mu_{U_0}(U_0) = 1$, and let μ_{J_0/U_0} be the point measure of the discrete group J_0/U_0. Let

$$\mu_{J_0} = \mu_{J_0/U_0} * \mu_{U_0}.$$

Let $\sigma : J_\infty \to \mathbf{R}^{*r_1} \times \mathbf{C}^{*r_2}, \sigma(\alpha) = (x_1, \ldots, x_r), r = r_1 + r_2$, be the topological isomorphism defined in §2.2. Let μ' be the Haar measure of $\mathbf{R}^{*r_1} \times \mathbf{C}^{*r_2}$ which is the product of the standard measures on the factors. Let

$$\mu_{J_\infty} \approx \mu',$$

under σ.

Let

$$\mu_J = \mu_{J_0} \times \mu_{J_\infty}.$$

Thus the Haar measure μ_J on J is fixed.

Let $\mu_{\mathbf{R}^+}$ be the standard measure on $\mathbf{R}^+$, and let

$$\mu_T \approx \mu_{\mathbf{R}^+}$$

under the isomorphism $T \to \mathbf{R}^+$, $\tau_x \mapsto x$. Then there exists a unique Haar measure μ_{J_1} on J_1 such that

$$\mu_J = \mu_{J_1} \times \mu_T.$$

Let μ_{F^*} be the point measure on the discrete group F^*, and put

$$\mu_{\overline{J}} = \mu_J/\mu_{F^*}, \quad \mu_{\overline{J_1}} = \mu_{J_1}/\mu_{F^*}.$$

Let

$$\mu_{\overline{T}} \approx \mu_T$$

under $\overline{T} = (F^* \times T)/F^* \to T$. Then

$$\mu_{\overline{J_1}} \times \mu_{\overline{T}} \approx \mu_{\overline{J_1}} \times \mu_T$$

under the isomorphism $\overline{J} = \overline{J_1} \times \overline{T} \to \overline{J_1} \times T$ induced by $\overline{T} \to T$. However, the above isomorphism can be written in the form

$$\overline{J} = (J_1 \times T)/(F^* \times 1) \to (J_1/F^*) \times (T/1).$$

Let μ_1 be the point measure of 1. Then $\mu_{F^*} \times \mu_1 = \mu_{F^*}$, $\mu_T/\mu_1 = \mu_T$. By 10 in the above, we have

$$(\mu_{J_1} \times \mu_T)/(\mu_{F^*} \times \mu_1) \approx (\mu_{J_1}/\mu_{F^*}) \times (\mu_T/\mu_1),$$

under $\overline{J} \to (J_1/F^*) \times (T/1)$, namely,

$$\mu_{\overline{J}} = \mu_J/\mu_{F^*} \approx \mu_{\overline{J_1}} \times \mu_T.$$

Therefore

$$\mu_{\overline{J}} = \mu_{\overline{J_1}} \times \mu_{\overline{T}}.$$

Remark 2.3 Let μ be any Haar measure of the additive group R. Let ξ be an element of J. Then $\sigma_\xi : R \to R$, $\alpha \mapsto \xi\alpha$, is a topological isomorphism. Let $\mu' = \mu^{\sigma_\xi^{-1}}$ so that $\mu'(A) = \mu(\sigma_\xi(A)) = \mu(\xi A)$ for any Borel set A in R. Then μ' is also a Haar measure of R, and $\mu' = c_\xi \mu$ with some constant $c_\xi > 0$. It can be proved that $c_\xi = V(\xi)$, namely, that

$$\mu(\xi A) = V(\xi)\mu(A), \quad \xi \in J, \ A \subset R.$$

This gives us an invariant definition of the volume $V(\xi)$.

Let $a = \xi$ be in F^*. Then $\sigma_a(F) = F$ so that σ_a induces an isomorphism $\sigma_a' : R/F \to R/F$. Let $\widetilde{\mu}$ be a Haar measure of R/F. Then we see easily that $\widetilde{\mu}(\sigma_a'(A')) = c_a\widetilde{\mu}(A')$ for any Borel set A' in R/F, with the same $c_a = V(a)$. Let $A' = R/F$. Then $\sigma_a'(A') = A'$. As R/F is compact, $0 < \widetilde{\mu}(A') < +\infty$. Hence it follows from the above that $c_a = 1$, namely, that $V(a) = 1, a \in F^*$. Thus the product formula for F is essentially a consequence of the compactness of R/F.

Chapter 3
L-functions

3.1 Definition

Let D be a domain in the complex s-plane. Let $u_1(s), u_2(s), \ldots,$ be a sequence of holomorphic functions on D such that $u_m(s) \neq 1$, $s \in D$, $m \geq 1$, and that $\sum_{m=1}^{\infty} |u_m(s)|$ converges uniformly on D to a bounded function on D. Then

$$\lim_{N\to\infty} \prod_{m=1}^{N} (1 - u_m(s)), \quad s \in D,$$

converges uniformly to a holomorphic function $u(s)$ on D such that $u(s) \neq 0, s \in D$. Furthermore, if $\{j_1, j_2, \ldots\}$ is any permutation of $\{1, 2, \ldots\}$, then $\lim_{N\to\infty} \prod_{m=1}^{N} (1 - u_{j_m}(s))$ converges to the same $u(s)$ in a similar way. The function $u(s)$ is denoted by

$$\prod_{m=1}^{\infty} (1 - u_m(s)), \quad s \in D.$$

Now, let F be as before a number field, and let $\chi : J \to \mathbf{C}_1$ be a Hecke character of F; $\chi(F^* \times T) = 1$. Let $\mathfrak{f}$ be the conductor of χ, and let $\chi(= \chi') : \mathfrak{I}_{\mathfrak{f}} \to \mathbf{C}_1$ be the character of the ideal group $\mathfrak{I}_{\mathfrak{f}}$ induced by $\chi : J \to \mathbf{C}_1$. For any integral ideal $\mathfrak{a}$ which is not prime to $\mathfrak{f}$, put

$$\chi(\mathfrak{a}) = 0.$$

Then $\chi(\mathfrak{a})$ is defined for every integral ideal of F, and

$$\chi(\mathfrak{a}\mathfrak{b}) = \chi(\mathfrak{a})\chi(\mathfrak{b})$$

for any integral ideal $\mathfrak{a}$ and $\mathfrak{b}$.

K. Iwasawa, *Hecke's L-functions*, SpringerBriefs in Mathematics,
https://doi.org/10.1007/978-981-13-9495-9_3

Let $\mathfrak{p}$ be a prime ideal of $\mathfrak{o} \subset F$. We write $\mathfrak{p}|p$ if $p(= P_p)$ is the restriction of $P_\mathfrak{p}$ on $\mathbf{Q}$. In such a case, $N(\mathfrak{p}) = p^f$, $f \geq 1$. Consider the function $N(\mathfrak{p})N(\mathfrak{p})^{-s}$ which is holomorphic on the entire s-plane. Since

$$|\chi(\mathfrak{p})N(\mathfrak{p})^{-s}| = |\chi(\mathfrak{p})||N(\mathfrak{p})^{-s}| \leq N(\mathfrak{p})^{-\mathrm{Re}\,(s)}$$

we have

$$|\chi(\mathfrak{p})N(\mathfrak{p})^{-s}| < 1, \quad \chi(\mathfrak{p})N(\mathfrak{p})^{-s} \neq 1, \quad \text{for } \mathrm{Re}\,(s) > 0.$$

Let

$$1 - u_\mathfrak{p}(s) = (1 - \chi(\mathfrak{p})N(\mathfrak{p})^{-s})^{-1},$$

namely,

$$u_\mathfrak{p}(s) = \frac{-\chi(\mathfrak{p})N(\mathfrak{p})^{-s}}{1 - \chi(\mathfrak{p})N(\mathfrak{p})^{-s}}, \quad \mathrm{Re}\,(s) > 0.$$

Then $u_\mathfrak{p}(s)$ is clearly holomorphic, and $u_\mathfrak{p}(s) \neq 1$ for $\mathrm{Re}\,(s) > 0$.

Let $a > 1$ and let $D_a = \{s \mid \mathrm{Re}\,(s) > a\}$. For s in D_a, $|x(\mathfrak{p})N(\mathfrak{p})| \leq N(\mathfrak{p})^{-\mathrm{Re}\,(s)} \leq N(\mathfrak{p})^{-a} = p^{-af} \leq p^{-a} \leq 2^{-a} \leq \frac{1}{2}$. Hence

$$|u_\mathfrak{p}(s)| \leq \frac{p^{-a}}{1 - \frac{1}{2}} = 2p^{-a}, \quad s \in D_a,$$

For each prime number p, there exist at most $n = [F : \mathbf{Q}]$ prime ideals $\mathfrak{p}$ such that $\mathfrak{p}|p$ (cf. I. §2). Hence, when $\mathfrak{p}$ ranges over all prime ideals of $\mathfrak{o}$, we obtain

$$\begin{aligned}\sum_\mathfrak{p} |u_\mathfrak{p}(s)| &= \sum_p \left(\sum_{\mathfrak{p}|p} |u_\mathfrak{p}(s)| \right) \leq \sum_p \left(\sum_{\mathfrak{p}|p} 2p^{-a} \right) \\ &\leq \sum_p 2np^{-a} = 2n \sum_p p^{-a} < +\infty, \quad s \in D_a.\end{aligned}$$

This shows that $\sum_\mathfrak{p} |u_\mathfrak{p}(s)|$ converges uniformly to a bounded function on D_a (and also that $\{u_\mathfrak{p}(s)\}$ is a countable set). Applying the result stated at the beginning to $D = D_a$, we obtain a holomorphic function

$$\prod_\mathfrak{p} (1 - u_\mathfrak{p}(s)) = \prod_\mathfrak{p} (1 - \chi(\mathfrak{p})N(\mathfrak{p})^{-s})^{-1}$$

on D_a. Since this holds for any $a > 1$, we see that the same product defines a function of s for any s with $\mathrm{Re}\,(s) > 1$. So, let

$$L(s; \chi) = \prod_\mathfrak{p} (1 - \chi(\mathfrak{p})N(\mathfrak{p})^{-s})^{-1}, \quad \text{for } \mathrm{Re}\,(s) > 1.$$

$L(s;\chi)$ is called Hecke's L-function for F with the character χ, as it follows from the above, $L(s,\chi)$ is holomorphic and $L(s;\chi) \neq 0$ for $\mathrm{Re}(s) > 1$, and

$$\prod_{k=1}^{m} (1 - \chi(\mathfrak{p}_k) N(\mathfrak{p}_k)^{-s})^{-1}$$

converges uniformly to $L(s;\chi)$ in $D_a = \{s \mid \mathrm{Re}(s) > a\}$, $a > 1$, for any ordering $\mathfrak{p}_1, \mathfrak{p}_2, \mathfrak{p}_3, \ldots,$ of the prime ideals of $\mathfrak{o}$.

Let $\chi \equiv 1$. Then $\mathfrak{f} = \mathfrak{o}$, and $\chi(\mathfrak{a}) = 1$ for any ideal $\mathfrak{a}$ of F. In this case, we write $\zeta_F(s)$ for $L(s;\chi)$;

$$\zeta_F(s) = \prod_{\mathfrak{p}} (1 - N(\mathfrak{p}^{-s})^{-1}, \quad \mathrm{Re}(s) > 1.$$

$\zeta_F(s)$ is called Dedekind's zeta-function for the number field F.

Proposition 3.1 *When $\mathfrak{a}$ ranges over all integral ideals of F, the Dirichlet series*

$$\sum_{\mathfrak{a}} \chi(\mathfrak{a}) N(\mathfrak{a})^{-s}$$

converges absolutely and uniformly on each domain D_a, $a > 1$, and

$$L(s;\chi) = \sum_{\mathfrak{a}} \chi(\mathfrak{a}) N(\mathfrak{a})^{-s}, \quad \textit{for } \mathrm{Re}(s) > 1.$$

Proof For any real $x > 0$, there exist only a finite number of prime ideals $\mathfrak{p}$ such that $N(\mathfrak{p}) \leq x$. We have

$$\begin{aligned} \prod_{N(\mathfrak{p}) \leq x} (1 - N(\mathfrak{p})^{-a})^{-1} &= \prod_{N(\mathfrak{p}) \leq x} (1 + N(\mathfrak{p})^{-a} + N(\mathfrak{p})^{-2a} + \cdots) \\ &= \sum_{\mathfrak{a}}{}' N(\mathfrak{a})^{-a}, \end{aligned}$$

where the sum is taken over all integral ideals $\mathfrak{a}$ such that the prime factors $\mathfrak{p}$ satisfy $N(\mathfrak{p}) \leq x$. Clearly every integral ideal $\mathfrak{a}$ with $N(\mathfrak{a}) \leq x$ is included in the above sum. Hence

$$\begin{aligned} \sum_{\substack{N(\mathfrak{a}) \leq x \\ \mathfrak{a} \subset \mathfrak{o}}} N(\mathfrak{a})^{-a} \leq \sum_{\mathfrak{a}}{}' N(\mathfrak{a})^{-a} &= \prod_{N(\mathfrak{p}) \leq x} (1 - N(\mathfrak{p})^{-a})^{-1} \\ &\leq \prod_{\mathfrak{p}} (1 - N(\mathfrak{p})^{-a})^{-1} = \zeta_F(a). \end{aligned}$$

Since this holds for any $x > 0$, it follows that

$$\sum_{\mathfrak{a}\subset\mathfrak{o}} N(\mathfrak{a})^{-a} \le \zeta_F(a) < +\infty.$$

Now, let s be in D_a. Since $|\chi(\mathfrak{a})N(\mathfrak{a})^{-s}| \le N(\mathfrak{a})^{-\mathrm{Re}(s)} \le N(\mathfrak{a})^{-a}$, we see from the above that $\sum \chi(\mathfrak{a})N(\mathfrak{a})^{-s}$ converges absolutely and uniformly in D_a. A computation similar to the above also shows that

$$\prod_{N(\mathfrak{p}\le x)} (1-\chi(\mathfrak{p})N(\mathfrak{p})^{-s})^{-1} = \sum_{\mathfrak{a}}{}' \chi(\mathfrak{a})N(\mathfrak{a})^{-s}, \quad s \in D_a,$$

because $\chi(\mathfrak{a}\mathfrak{b}) = \chi(\mathfrak{a})\chi(\mathfrak{b})$, $\mathfrak{a}, \mathfrak{b} \subset \mathfrak{o}$. Hence

$$\begin{aligned}\left|\prod_{N(\mathfrak{p})\le x} (1-\chi(\mathfrak{p})N(\mathfrak{p})^{-s})^{-1} - \sum_{N(\mathfrak{a})\le x} \chi(\mathfrak{a})N(\mathfrak{a})^{-s}\right| &= \left|\sum_{N(\mathfrak{a})>x}{}' N(\mathfrak{a})^{-s}\right| \\ &\le \sum_{N(\mathfrak{a})>x}{}' N(\mathfrak{a})^{-a} \le \sum_{N(\mathfrak{a})>x} N(\mathfrak{a})^{-a}.\end{aligned}$$

Let $x \to +\infty$. Since $\sum_{\mathfrak{a}} N(\mathfrak{a})^{-a}$ converges, $\sum_{N(\mathfrak{a})>x} N(\mathfrak{a})^{-a} \to 0$. Hence we obtain

$$L(s;\chi) = \prod_{\mathfrak{p}}(1-\chi(\mathfrak{p})N(\mathfrak{p})^{-s})^{-1} = \sum_{\mathfrak{a}} \chi(\mathfrak{a})N(\mathfrak{a})^{-s}, \quad s \in D_a.$$

Since $a(>1)$ is arbitrary, the above equality also holds for any s with $\mathrm{Re}(s) > 1$.

For $\chi \equiv 1$, we have

$$\zeta_F(s) = \prod_{\mathfrak{p}}(1-N(\mathfrak{p})^{-s})^{-1} = \sum_{\mathfrak{a}} N(\mathfrak{a})^{-s}, \quad \mathrm{Re}(s) > 1.$$

Let $F = \mathbf{Q}$. Then $\zeta(s) = \zeta_{\mathbf{Q}}(s)$ is the well-known zeta-function of Riemann;

$$\zeta(s) = \prod_{p}(1-p^{-s})^{-1} = \sum_{m=1}^{\infty} m^{-s}, \quad \mathrm{Re}(s) > 1.$$

For $\zeta(s)$, Riemann proved the following. Let $s > 1$. We have

$$\Gamma\left(\frac{s}{2}\right) = \int_0^{\infty} e^{-x}x^{\frac{s}{2}}\frac{dx}{x}.$$

Hence

$$\pi^{-\frac{s}{2}}\Gamma\left(\frac{s}{2}\right)m^{-s} = \int_0^\infty e^{-\pi m^2 x}x^{\frac{s}{2}}\frac{dx}{x}.$$

It follows that

$$\begin{aligned}\pi^{-\frac{s}{2}}\Gamma\left(\frac{s}{2}\right)\zeta(s) &= \sum_{m=1}^{\infty}\int_0^\infty e^{-\pi m^2 x}x^{\frac{s}{2}}\frac{dx}{x}\\ &= \int_0^\infty \sum_{m=1}^{\infty} e^{-\pi m^2 x}x^{\frac{s}{2}}\frac{dx}{x}\\ &= \int_0^\infty \omega(x)x^{\frac{s}{2}}\frac{dx}{x},\end{aligned}$$

where

$$\omega(x) = \sum_{m=1}^{\infty} e^{-\pi m^2 x}, \quad x > 0.$$

The interchange of the sum and the integral is allowed because $e^{-\pi m^2 x}x^{\frac{s}{2}} \geq 0$. Let

$$\theta(x) = \sum_{m\in\mathbf{Z}} e^{-\pi m^2 x} = 1 + 2\omega(x), \quad x > 0.$$

Then the classical theta-formula states that

$$\theta(x) = \frac{1}{\sqrt{x}}\,\theta\left(\frac{1}{x}\right), \quad x > 0,$$

namely, that

$$\omega(x) = \frac{1}{\sqrt{x}}\,\omega\left(\frac{1}{x}\right) - \frac{1}{2} + \frac{1}{2\sqrt{x}}, \quad x > 0.$$

Hence we have

$$\int_0^\infty \omega(x)x^{\frac{s}{2}}\frac{dx}{x} = \int_0^1 \omega(x)x^{\frac{s}{2}}\frac{dx}{x} + \int_1^\infty \omega(x)x^{\frac{s}{2}}\frac{dx}{x},$$

$$\int_0^1 \omega(x)x^{\frac{s}{2}}\frac{dx}{x} = \int_0^1 \left(\frac{1}{\sqrt{x}}\,\omega\left(\frac{1}{x}\right) - \frac{1}{2} + \frac{1}{2\sqrt{x}}\right)x^{\frac{s}{2}}\frac{dx}{x},$$

$$\begin{aligned}\int_0^1 \left(-\frac{1}{2}x^{\frac{s}{2}} + \frac{1}{2}x^{\frac{s-1}{2}}\right)\frac{dx}{x} &= \left[-\frac{1}{2}\frac{x^{\frac{s}{2}}}{\frac{s}{2}} + \frac{1}{2}\frac{x^{\frac{s-1}{2}}}{\frac{s-1}{2}}\right]_0^1, \quad s > 1\\ &= -\frac{1}{s} + \frac{1}{s-1}\\ &= \frac{1}{s(s-1)},\end{aligned}$$

so that

$$\int_0^1 \omega(x)x^{\frac{s}{2}}\frac{dx}{x} = \int_0^1 \omega\left(\frac{1}{x}\right)x^{\frac{s-1}{2}}\frac{dx}{x} + \frac{1}{s(s-1)}$$
$$= \int_1^\infty \omega(x)x^{\frac{1-s}{2}}\frac{dx}{x} + \frac{1}{s(s-1)},$$
$$\pi^{-\frac{s}{2}}\Gamma\left(\frac{s}{2}\right)\zeta(s) = \int_1^\infty \omega(x)x^{\frac{s}{2}}\frac{dx}{x} + \int_1^\infty \omega(x)x^{\frac{1-s}{2}}\frac{dx}{x} + \frac{1}{s(s-1)}, \quad s > 1.$$

This shows in particular that

$$\int_1^\infty \omega(x)x^{\frac{s}{2}}\frac{dx}{x} < +\infty$$

for any $s > 1$.

Let $a > 1$, and let s be any complex number such that $\mathrm{Re}\,(s) < a$. Then $|x^{\frac{s}{2}}| = x^{\mathrm{Re}\,(\frac{s}{2})} \leq x^{\frac{a}{2}}$ for $x \geq 1$. Hence

$$\int_1^\infty |\omega(x)x^{\frac{s}{2}}|\frac{dx}{x} \leq \int_1^\infty \omega(x)x^{\frac{a}{2}}\frac{dx}{x} < +\infty.$$

This shows that

$$\eta(s) = \int_1^\infty \omega(x)x^{\frac{s}{2}}\frac{dx}{x}$$

defines a holomorphic function of s for $\mathrm{Re}\,(s) < a$. Since a (> 1) is arbitrary, $\eta(s)$ actually defines a holomorphic function on the entire s-plane.

Let

$$\xi(s) = \pi^{-\frac{s}{2}}\Gamma\left(\frac{s}{2}\right)\zeta(s), \quad \mathrm{Re}\,(s) > 1.$$

It follows from the above that

$$\xi(s) = \eta(s) + \eta(1-s) + \frac{1}{s(s-1)}, \quad \text{for } s > 1. \tag{3.1}$$

Since both sides of the above are holomorphic function of s for $\mathrm{Re}\,(s) > 1$, the equality holds for any s with $\mathrm{Re}\,(s) > 1$. Moreover, the right-hand side is a meromorphic function on the entire s-plane with the only simple poles at $s = 0, 1$. Hence, by the above equality $\xi(s)$ can be continuous analytically on the entire s-plane, and so is $\zeta(s) = \pi^{\frac{s}{2}}\Gamma\left(\frac{s}{2}\right)^{-1}\xi(s)$. Since $\Gamma\left(\frac{s}{2}\right)^{-1}$ has a zero at $s = 0$, $\zeta(s)$ is regular at $s = 0$. Since $\Gamma\left(\frac{1}{2}\right) = \pi^{\frac{1}{2}}$ and since $\xi(s)$ has a simple pole with residue 1 at $s = 1$, $\zeta(s)$ also has a simple pole with residue 1 at $s = 1$. Furthermore, it follows from (3.1) (which now holds for arbitrary s) that

$$\xi(s) = \xi(1 - s).$$

This is called the functional equation for $\zeta(s)$.

The original proof of Riemann for these results is somewhat different from what is stated in the above. In the following, we shall prove similar results for Hecke's L-functions $L(s; \chi)$, generalizing the method used above.

3.2 Theta-Formulae (Analytic Form)

Let $n \geq 1$. For $x = (x_1, \ldots, x_n)$, $y = (y_1, \ldots, y_n)$ in $\mathbf{C}^n$, let

$$(x, y) = \sum_{k=1}^{n} x_k y_k, \quad \|x\| = (x, \overline{x})^{\frac{1}{2}} = \left(\sum_{k=1}^{n} |x_k|^2\right)^{\frac{1}{2}}.$$

Let A be an $n \times n$, real, positive definite and symmetric matrix, and put

$$A(x, y) = (xA, y) = (x, y\,{}^*A) = \sum_{j,k=1}^{n} a_{jk} x_j y_k,$$

$$A(x) := A(x, x), \quad x, y \in \mathbf{C}^n.$$

Since A is symmetric, $A(x, y) = A(y, x)$; and since A is real and positive definite, there exists an $a > 0$ such that

$$A(x) \geq a\|x\|^2, \quad \text{for any } x \in \mathbf{R}^n.$$

For $x, y \in \mathbf{C}^n$, let

$$\theta(x, y; A) = \sum_{m} \exp(-\pi A(x + m) + 2\pi i(y, m)),$$

where $m = (m_1, \ldots, m_n)$ ranges over all integral vectors in $\mathbf{Z}^n$.

Lemma 3.1 *The above series converges absolutely and uniformly in x and y when x and y remain in a bounded domain D of $\mathbf{C}^n$. Hence $\theta(x, y; A)$ is a holomorphic function of $(x_1, \ldots, x_n, y_1, \ldots, y_n)$ in $\mathbf{C}^{2n}$.*

Proof Let

$$q(m; x, y) = -\pi A(x + m) + 2\pi i(y, m).$$

Since $A(x + m) = A(m) + 2A(x, m) + A(x)$, $A(x, m) = (xA, m)$,

$$q(m; x, y) = -\pi A(m) + (u, m) - \pi A(x),$$

where $u = -2\pi x A + 2\pi i y$. For the given bounded domain D in $\mathbf{C}^n$, there exists a constant $C > 0$ such that

$$\|u\| \le C, \quad |\pi A(x)| \le C, \quad x, y \in D.$$

Let $m = (m_1, \ldots, m_n) \in \mathbf{Z}^n$, $\|m\| < \frac{2c}{a\pi}$. Then $|m_k| < \frac{2c}{a\pi}$, $m_k \in \mathbf{Z}$. Hence there exist only a finite number of such m's in $\mathbf{Z}^n$. Therefore it is sufficient to show that

$$\sum_{\|m\| \ge \frac{2C}{a\pi}} \exp(q(m; x, y))$$

converges absolutely and uniformly for x and y in D. For such an m,

$$|(u, m)| + |\pi A(x)| \le \|u\| \|m\| + C \le C\|m\| + C \le \frac{a\pi}{2} \|m\|^2 + C,$$

so that

$$\begin{aligned} \mathrm{Re}\,(q(m; x, y)) &= -\pi A(m) + \mathrm{Re}\,((u, m) - \pi A(x)) \\ &\le -\pi a \|m\|^2 + \frac{a\pi}{2} \|m\|^2 + C \\ &= -\frac{\pi a}{2} \|m\|^2 + C. \end{aligned}$$

Hence

$$\begin{aligned} \sum_{\|m\| \ge \frac{2c}{a\pi}} |\exp(q(m; x, y))| &= \sum_m \exp(\mathrm{Re}\,(q(m; x, y))) \\ &\le C' \sum_m \exp\left(-\frac{\pi a}{2} \|m\|^2\right), \quad C' = e^C \\ &\le C' \sum_{m \in \mathbf{Z}^n} \exp\left(-\frac{\pi a}{2} \|m\|^2\right) \\ &= C' \left(\sum_{l \in \mathbf{Z}} \exp\left(-\frac{\pi a}{2} l^2\right)\right)^2 \\ &= C' \left(1 + 2\sum_{l=1}^{\infty} \exp\left(-\frac{\pi a}{2} l^2\right)\right)^n \\ &\le C' \left(1 + 2\sum_{l=1}^{\infty} \exp\left(-\frac{\pi a}{2} l\right)\right) \\ &= C' \left(1 + 2\frac{e^{-\frac{\pi a}{2}}}{1 - e^{-\frac{\pi a}{2}}}\right)^n. \end{aligned}$$

This proves our assertion.

Lemma 3.2 $\theta(x, 0; A) = \frac{1}{\sqrt{|A|}}\theta(0, x; A^{-1})$, $x \in \mathbf{C}^n$.

Proof We first note that if A is real, positive definite and symmetric, then so is A^{-1}, and $|A| > 0$. Hence $\theta(x, y; A^{-1})$ is defined, and is a holomorphic function of x and y. Since both sides of the above are holomorphic in $x \in \mathbf{C}^n$, it is sufficient to prove the equality for real $x \in \mathbf{R}^n$. Let

$$f(x) = \theta(x, 0; A) = \sum_m \exp(-\pi A(x + m)), \quad x \in \mathbf{R}^n.$$

Then $f(x + m') = f(x)$ for any m' in $\mathbf{Z}^n$, namely, $f(x)$ is periodic in each x_k with period 1. Since $f(x)$ is of class C^∞ on $\mathbf{R}^n$, it follows from the standard theorem on Fourier series that

$$f(x) = \sum_{m \in \mathbf{Z}^n} a(m) \exp(2\pi i(x, m)),$$

where the right-hand side converges absolutely and uniformly on each bounded domain of $\mathbf{R}^n$, and where $a(m)$ is given by

$$a(m) = \int_0^1 \cdots \int_0^1 f(x) \exp(-2\pi i(x, m)) dx_1 \cdots dx_n.$$

Now, the above integral is equal to

$$\begin{aligned}
&\int_0^1 \cdots \int_0^1 \sum_{m'} \exp\left(-\pi A(x + m') - 2\pi i(x, m)\right) dx \\
&= \sum_{m'} \int_0^1 \cdots \int_0^1 \exp\left(-\pi A(x + m') - 2\pi i(x + m', m)\right) dx \\
&= \sum_{m'} \int_{m'_1}^{m'_1+1} \cdots \int_{m'_n}^{m'_n+1} \exp\left(-\pi A(x) - 2\pi i(x, m)\right) dx \\
&= \int_{-\infty}^{\infty} \cdots \int_{-\infty}^{\infty} \exp\left(-\pi A(x) - 2\pi i(x, m)\right) dx_1 \cdots dx_n.
\end{aligned}$$

Since A is positive definite, there exists an $n \times n$, real, positive definite and symmetric matrix B such that $A = B^2$. Let

$$y = xB, \quad t = mB^{-1}.$$

Then $A(x) = (xA, x) = (xB^2, x) = (xB, xB) = (y, y)$, $(x, m) = (yB^{-1}, m) = (y, mB^{-1}) = (y, t)$, so that $A(x) + 2i(x, m) = (y, y) + 2(y, it) = (y + it, y + it) + (t, t)$, where $(t, t) = (mB^{-1}, mB^{-1}) = (mB^{-2}, m) = (mA^{-1}, m) = A^{-1}(m)$. Since

$$dy_1 \cdots dy_n = |B| dx_1 \cdots dx_n = \sqrt{|A|} dx_1 \cdots dx_n,$$

we see that the above integral is equal to

$$\frac{1}{\sqrt{|A|}}\exp\left(-\pi A^{-1}(m)\right)\int_{-\infty}^{\infty}\cdots\int_{-\infty}^{\infty}\exp\left(-\pi\sum_{k=1}^{n}(y_k+it_k)^2\right)dy_1\cdots dy_n.$$

However, one sees easily that

$$\int_{-\infty}^{\infty}\exp\left(-\pi(z+u)^2\right)dz=\int_{-\infty}^{\infty}\exp\left(-\pi z^2\right)dz=1.$$

Therefore

$$a(m)=\frac{1}{\sqrt{|A|}}\exp\left(-\pi A^{-1}(m)\right).$$

It follows that

$$\begin{aligned}\theta(x,0;A)=f(x)&=\sum_m\frac{1}{\sqrt{|A|}}\exp\left(-\pi A^{-1}(m)+2\pi i(x,m)\right)\\&=\frac{1}{\sqrt{|A|}}\theta(0,x;A^{-1}).\end{aligned}$$

Proposition 3.2 *For any x, y in $\mathbf{C}^n$,*

$$\theta(x,y;A)=\frac{1}{\sqrt{|A|}}\exp\left(-2\pi i(x,y)\right)\theta(-y,x;A^{-1}).$$

Proof In Lemma 3.2, we make the substitution $x\mapsto x-iyA^{-1}$. Then the exponent of the m-term on the left is

$$-\pi A(x-iyA^{-1}+m)=-\pi A(x+m)-2\pi A(x+m,-iyA^{-1})-\pi A(-iyA^{-1}).$$

Here

$$\begin{aligned}A(x+m,-iyA^{-1})&=(x+m,-iy)=-i(x+m,y)\\A(-iyA^{-1})&=(-iyA^{-1},-iy)=-(yA^{-1},y)=-A^{-1}(y).\end{aligned}$$

Hence

$$-\pi A(x-iyA^{-1}+m)=-\pi A(x+m)+2\pi i(y,m)+2\pi i(x,y)+\pi A^{-1}(y).$$

The exponent of of the corresponding term on the right is

$$-\pi A^{-1}(m)+2\pi i(x-iyA^{-1},m)=-\pi A^{-1}(m)+2\pi i(x,m)+2\pi A^{-1}(y,m).$$

Since

$$-\pi A^{-1}(m) + 2\pi A^{-1}(y, m) - \pi A^{-1}(y) = -\pi A^{-1}(-y + m),$$

we obtain the formula stated in the proposition.

Remark 3.1 Let $y = 0$ in the above. Then we have the equality in Lemma 3.2. Let also $x = 0$. Then

$$\theta(0, 0; A) = \frac{1}{\sqrt{|A|}}\theta(0, 0; A^{-1}),$$

namely,

$$\sum_m \exp(-\pi A(m)) = \frac{1}{\sqrt{|A|}} \sum_m \exp(-\pi A^{-1}(m)).$$

Let $n = 1$ and $A = (x)$, $A(m) = xm^2$, $x > 0$. Then we obtain

$$\sum_{m \in \mathbf{Z}} \exp(-\pi m^2 x) = \frac{1}{\sqrt{x}} \sum_{m \in \mathbf{Z}} \exp\left(-\pi m^2 \frac{1}{x}\right),$$

namely, the formula $\theta(x) = \frac{1}{\sqrt{x}}\theta(\frac{1}{x})$ stated in Sect. 3.1.

3.3 Theta-Formulae (Arithmetic Form)

Let F be a number field, $n = [F : \mathbf{Q}]$, and let r_1, r_2 be as before; $n = r_1 + 2r_2$. Let $c_1, \ldots, c_n$ be positive reals such that $c_j = c_{j+r_2}$, $r_1 + 1 \le j \le r_1 + r_2$. Put $C = (\delta_{jk} c_k)$. Let $\mathfrak{a}$ be a non-zero ideal of F, $\alpha = (a_1, \ldots, a_n)$ a basis of $\mathfrak{a}$ over $\mathbf{Z}$, and $M_\alpha = (a_j^{(k)})$. Put

$$A = M_\alpha C\, {}^t\overline{M}_\alpha.$$

A is an $n \times n$ matrix whose (j, k)-entry is $\sum_{l=1}^n a_j^{(l)} c_l \overline{a_k}^{(l)}$. Using $c_j = c_{j+r_2}$, $r_1 + 1 \le j \le r_1 + r_2$, we see easily that A is real and symmetric. However, it can be proved also as follows. For any $x = (x_1, \ldots, x_n)$ in $\mathbf{C}^n$, let

$$\begin{aligned} u(x) &= xM_\alpha \\ &= (u_1(x), \ldots, u_n(x)), \\ u_k(x) &= \sum_{j=1}^n a_j^{(k)} x_j, \quad k = 1, \ldots, n. \end{aligned}$$

Since $\overline{a}_j^{(k)} = a_j^{(k)}$, $1 \le k \le r_1$, $\overline{a}_j^{(k)} = a_j^{(k+r_2)}$, $r_1 + 1 \le k \le r_1 + r_2$, we have

$$
\begin{aligned}
\overline{u_k(x)} &= u_k(\overline{x}), \quad 1 \le k \le r_1, \\
\overline{u_k(x)} &= u_{k+r_2}(\overline{x}), \\
\overline{u_{k+r_2}(x)} &= u_k(\overline{x}), \quad r_1 + 1 \le k \le r_1 + r_2.
\end{aligned}
$$

As $c_k = c_{k+r_2}$, $r_1 + 1 \le k \le r_1 + r_2$, it follows that

$$
\begin{aligned}
A(x, y) &= xA^t y = xM_\alpha C\,{}^t\overline{(\overline{y}M_\alpha)} = u(x)C\,{}^t\overline{u(\overline{y})} \\
&= \sum_{k=1}^{r_1} c_k u_k(x)u_k(y) + \sum_{k=r_1+1}^{r_1+r_2} c_k \left(u_k(x)u_{k+r_2}(y) + u_{k+r_2}(x)u_k(y)\right).
\end{aligned}
$$

Hence $A(x, y) = A(y, x)$ for any x, y in $\mathbf{C}^n$, and A is therefore symmetric. Let $x = y \in \mathbf{R}^n$ in the above. Then

$$
A(x) = \sum_{k=1}^{n} c_k |u_k(x)|^2, \quad x \in \mathbf{R}^n. \tag{3.2}
$$

Hence $A(x) \ge 0$. As A is symmetric, we see that A is real. Furthermore, $A(x) = 0$ implies $u_k(x) = 0, k = 1, \ldots, n$, namely, $xM_\alpha = 0$. Since $|M_\alpha| \ne 0$, we have $x = 0$. Therefore A is positive definite. Thus A is an $n \times n$, real, symmetric and positive definite matrix.

We now write down the formula $\theta(x, 0; A) = \frac{1}{\sqrt{|A|}}\theta(0, x; A^{-1})$ for the above A. Let $m = (m_1, \ldots, m_n) \in \mathbf{Z}^n$, and let

$$
a = \sum_{k=1}^{n} a_k m_k.
$$

Since $\alpha = (a_1, \ldots, a_n)$ is a basis of $\mathfrak{a}$, a ranges over $\mathfrak{a}$ when m ranges over $\mathbf{Z}^n$. Also $u_j(m) = \sum_{k=1}^n a_k^{(j)} m_k = a^{(j)}$. Hence

$$
\begin{aligned}
u(m) &= (a^{(1)}, \ldots, a^{(n)}), \\
u(x + m) &= (u_1 + a^{(1)}, \ldots, u_n + a^{(n)}), \quad x \in \mathbf{C}^n,
\end{aligned}
$$

where $u_j = u_j(x)$. Therefore

$$
\begin{aligned}
&\theta(x, 0; A) \\
&= \sum_{a \in \mathfrak{a}} \exp\left(-\pi\left(\sum_{k=1}^{r_1} c_k(u_k + a^{(k)})^2 + 2\sum_{k=r_1+1}^{r_1+r_2} c_k(u_k + a^{(k)})(u_{k+r_2} + a^{(k+r_2)})\right)\right).
\end{aligned}
$$

On the other hand, we know that $(\det M_\alpha)^2 = \Delta N(\mathfrak{a})^2$ (cf. Sect. 1.1). Hence $\det A = (\det C)\,|\det M_\alpha|^2 = c_1 \cdots c_n d N(\mathfrak{a})^2$, where $d = |\Delta| = N(\mathfrak{d})$. Let $\beta = (b_1, \ldots, b_n)$ be a basis of $F/\mathbf{Q}$, complementary to α. Then β is a basis of the ideal $\widetilde{\mathfrak{a}} = \mathfrak{a}^{-1}\mathfrak{d}^{-1}$

over $\mathbf{Z}$, and $M_\beta = {}^t M_\alpha^{-1}$. It follows that

$$A^{-1} = \overline{A}^{-1} = {}^t M_\alpha^{-1} C^{-1} \overline{M_\alpha}^{-1} = M_\beta C^{-1} {}^t \overline{M}_\beta.$$

Let

$$v(x) = x M_\beta, \quad x = (x_1, \ldots, x_n) \in \mathbf{C}^n.$$

Then

$$A^{-1}(x) = \sum_{k=1}^{n} c_k^{-1} |v_k(x)|^2, \quad x \in \mathbf{R}^n,$$

just as in (3.2). When $m = (m_1, \ldots, m_n)$ ranges over $\mathbf{Z}^n$, $b = \sum_{k=1}^n b_k m_k$ ranges over $\widetilde{\mathfrak{a}} = \mathfrak{a}^{-1}\mathfrak{d}^{-1}$. Since $v(m) = (b^{(1)}, \ldots, b^{(n)})$, $A^{-1}(m) = \sum_{k=1}^n c_k^{-1} |b^{(k)}|^2$. Furthermore, $(x, m) = (x M_\alpha, m\, {}^t M_\alpha^{-1}) = (u(x), v(m)) = \sum_{k=1}^n u_k b^{(k)}$. Therefore

$$\theta(0, x; A^{-1}) = \sum_{b \in \mathfrak{a}^{-1}\mathfrak{d}^{-1}} \exp\left(-\pi \sum_{k=1}^{n} c_k^{-1} |b^{(k)}|^2 + 2\pi i \sum_{k=1}^{n} b^{(k)} u_k\right),$$

and we have

$$\sum_{a \in \mathfrak{a}} \exp\left(-\pi\left(\sum_{k=1}^{r_1} c_k (u_k + a^{(k)})^2 + 2 \sum_{k=r_1+1}^{r_1+r_2} c_k (u_k + a^{(k)})(u_{k+r_2} + a^{(k+r_2)})\right)\right)$$
$$= \frac{1}{\sqrt{c_1 \cdots c_n d N(\mathfrak{a})^2}} \sum_{b \in \mathfrak{a}^{-1}\mathfrak{d}^{-1}} \exp\left(-\pi \sum_{k=1}^{n} c_k^{-1} |b^{(k)}|^2 + 2\pi i \sum_{k=1}^{n} b^{(k)} u_k\right). \quad (3.3)$$

The equality holds for any $u = u(x) = x M_\alpha$, $x \in \mathbf{C}^n$, and hence also for arbitrary $u = (u_1, \ldots, u_n)$ in $\mathbf{C}^n$.

Let $e_1, \ldots, e_n$ be rational integers such that $e_k \geq 0$, $1 \leq k \leq n$, and

$$e_k = 0 \text{ or } 1, \quad 1 \leq k \leq r_1,$$
$$e_k e_{k+r_2} = 0, \quad r_1 + 1 \leq k \leq r_1 + r_2.$$

Let

$$e'_k = e_k, \quad 1 \leq k \leq r_1,$$
$$e'_k = e_{k+r_2},\ e'_{k+r_2} = e_k, \quad r_1 + 1 \leq k \leq r_1 + r_2.$$

We differentiate the both sides of (3.3) e_k-times in each variable u_k, $1 \leq k \leq n$; we may differentiate them term by term because the both series are absolutely and uniformly convergent in u in a bounded domain of $\mathbf{C}^n$. Hence we have

$$\sum_{a\in\mathfrak{a}}\prod_{k=1}^{n}(-2\pi c_k(u_k+a^{(k)}))^{e'_k}\exp(-\pi(\cdots))$$
$$=\frac{1}{\sqrt{c_1\cdots c_n dN((a))^2}}\sum_{b\in\mathfrak{a}^{-1}\mathfrak{d}^{-1}}\prod_{k=1}^{n}(2\pi i b^{(k)})^{e_k}\exp(\cdots),$$

namely

$$\prod_{k=1}^{n}c_k^{e'_k}\sum_{a\in\mathfrak{a}}\prod_{k=1}^{n}(u_k+a^{(k)})^{e'_k}\exp(-\pi(\cdots))$$
$$=\frac{(-i)^e}{\sqrt{c_1\cdots c_n dN(\mathfrak{a})^2}}\sum_{b\in\mathfrak{a}^{-1}\mathfrak{d}^{-1}}\prod_{k=1}^{n}(b^{(k)})^{e_k}\exp(\cdots),$$

where $e=\sum_{k=1}^{n}e_k=\sum_{k=1}^{n}e'_k$.

Let r be any element of F. Put $u_k=r^{(k)}$ in the above. Then we obtain the following result.

Proposition 3.3 *Let $\mathfrak{a}$ be a non-zero ideal of F, and let r be an element of F. Then*

$$\prod_{k=1}^{n}c_k^{e'_k}\sum_{a\in\mathfrak{a}}\prod_{k=1}^{n}(r^{(k)}+a^{(k)})^{e'_k}\exp\left(-\pi\sum_{k=1}^{n}c_k|r^{(k)}+a^{(k)}|^2\right)$$
$$=\frac{(-i)^e}{\sqrt{c_1\cdots c_n dN(\mathfrak{a})^2}}\sum_{b\in\mathfrak{a}^{-1}\mathfrak{d}^{-1}}\prod_{k=1}^{n}(b^{(k)})^{e_k}\exp\left(-\pi\sum_{k=1}^{n}c_k^{-1}|b^{(k)}|^2\right)+2\pi i T(rb).$$

Here $c_1,\ldots,c_n>0$, $c_j=c_{j+r_2}$ for $r_1+1\le j\le r_1+r_2$, $e_1,\ldots e_n$ and $e'_1,\ldots,e'_n$, are integers as stated in the above, and $e=\sum_{k=1}^{n}e_k=\sum_{k=1}^{n}e'_k$. Furthermore, the series on the both sides of the equality are absolutely convergent.

Now, let χ be a Hecke character of F, $\mathfrak{f}$ the conductor of χ, $f=N(\mathfrak{f})$, and $\{m_1,\ldots,m_n;s_1,\ldots s_n\}$ the signature of χ. Then the conjugate $\overline{\chi}$ of χ has the same conductor $\mathfrak{f}$ and the signature $\{m'_1,\ldots,m'_n;-s_1,\ldots,-s_n\}$ where

$$m'_k=m_k,\qquad 1\le k\le r_1,$$
$$m'_k=m_{k+r_2},\qquad m'_{k+r_2}=m_k,\qquad r_1+1\le k\le r_1+r_2.$$

Let $\alpha=(a_P)$ be any adèle of F; $\alpha\in R$, and let $t_k=\sigma_k(a_{P_{\infty,k}})$, $1\le k\le r$, $t_{k+r_2}=\overline{t}_k$, $r_1+1\le k\le r=r_1+r_2$. Let

$$g(\alpha;\chi)=\overline{\chi}_{\mathfrak{f}}(\alpha)\prod_{k=1}^{n}t_k^{m'_k}\exp\left(-\frac{\pi}{\sqrt[n]{df}}\sum_{k=1}^{n}|t_k|^2\right),\qquad \alpha\in R.$$

For any idèle α of F ($\alpha\in J$), we then define

$$\theta(\alpha;\chi)=\sum_{a\in\mathfrak{a}^{-1}} g(\alpha a;\chi),\quad \mathfrak{a}=\iota(\alpha). \tag{3.4}$$

When a runs over $\mathfrak{a}^{-1}\mathfrak{f}$, and r over a set of representatives of $\mathfrak{a}^{-1}$ modulo $\mathfrak{a}^{-1}\mathfrak{f}$, then $r+a$ obviously runs over all elements of $\mathfrak{a}^{-1}$ (once over each element). Hence

$$\theta(\alpha;\chi)=\sum_{r}\sum_{a\in\mathfrak{a}^{-1}\mathfrak{f}} g(\alpha(r+a);\chi).$$

Since $\alpha a,\alpha(r+a)\in R'$, $\alpha a=\alpha(a+r) \bmod R_{\mathfrak{f}}$, we have $\chi_{\mathfrak{f}}(\alpha r)=\chi_{\mathfrak{f}}(\alpha(r+a))$. Hence

$$g(\alpha(a+r))=\overline{\chi}_{\mathfrak{f}}(\alpha r)\prod t_k^{m'_k}\prod(r^{(k)}+a^{(k)})^{m'_k}\exp\left(-\frac{\pi}{\sqrt[n]{df}}\sum_{k=1}^{n}|t_k|^2|r^{(k)}+a^{(a)}|^2\right),$$

and consequently,

$$\theta(\alpha;\chi)=\prod t_k^{m'_k}\sum_{r}\overline{\chi}_{\mathfrak{f}}(\alpha r)\sum_{a\in\mathfrak{a}^{-1}\mathfrak{f}}\prod(r^{(k)}+a^{(k)})^{m'_k}\exp(\cdots).$$

If we replace $\mathfrak{a}$ by $\mathfrak{a}^{-1}\mathfrak{f}$, and put

$$c_k=\frac{|t_k|^2}{\sqrt[n]{df}},\quad e_k=m_k,\quad e'_k=m'_k,\qquad 1\le k\le n,$$

in Proposition 3.3, we see that the series $\sum_{a\in\mathfrak{a}^{-1}\mathfrak{f}}$ is absolutely convergent. Since r ranges over a finite set, it follows that the right-hand side of (3.4) is also absolutely convergent, and hence $\theta(\alpha;\chi)$ is well-defined. Furthermore, the same proposition shows that

$$\begin{aligned}\theta(\alpha;\chi)&=\prod t_k^{m'_k}\sum_{r}\overline{\chi}_{\mathfrak{f}}(\alpha r)\left(\prod c_k^{m'_k}\right)^{-1}\frac{(-i)^M}{\sqrt{c_1\cdots c_n d N(\mathfrak{a}^{-1}\mathfrak{f})^2}}\\&\quad\times\sum_{b\in\mathfrak{a}\mathfrak{d}^{-1}\mathfrak{f}^{-1}}\prod(b^{(k)})^{m_k}\exp\left(-\pi\sum_{k=1}^{n}\sigma_k^{-1}|b^{(k)}|^2+2\pi i T(rb)\right)\\&=h\sum_{b\in\mathfrak{a}\mathfrak{d}^{-1}\mathfrak{f}^{-1}}\prod\left(\frac{\sqrt[n]{df}\,b^{(k)}}{t_k}\right)^{m_k}\exp\left(-\frac{\pi}{\sqrt[n]{df}}\sum_{k=1}^{n}\left|\frac{\sqrt[n]{df}\,b^{(k)}}{t_k}\right|^2\right)\\&\quad\times\sum_{r}\overline{\chi}_{\mathfrak{f}}(\alpha r)\omega_0(rb),\end{aligned}$$

where

$$h = \frac{(-i)^M}{\sqrt{c_1 \cdots c_n d N(\mathfrak{a}^{-1}\mathfrak{f})^2}} \prod t_k^{m'_k} \left(\prod c_k^{m'_k}\right)^{-1} \prod \left(\frac{t_k}{\sqrt[n]{df}}\right)^{m_k}, \quad M = \sum_{k=1}^{n} m_k,$$
$$\omega_0(rb) = \exp(2\pi i T(rb)).$$

However, by Proposition 2.5, we know

$$\sum_r \overline{\chi}_{\mathfrak{f}}(\alpha r)\omega_0(\alpha r)\omega_0(rb) = \chi_{\mathfrak{f}}(\alpha^{-1}\delta\varphi b)C(\chi; \delta, \varphi),$$

for any b in $\mathfrak{a}\mathfrak{d}^{-1}\mathfrak{f}^{-1}$. Hence

$$\begin{aligned}\theta(\alpha; \chi) &= C(\chi; \delta, \varphi)\, h \sum_{b \in \mathfrak{a}\mathfrak{d}^{-1}\mathfrak{f}^{-1}} g(\alpha^{-1}\delta\varphi b; \overline{\chi}), \quad \iota(\alpha^{-1}\delta\varphi) = \mathfrak{a}^{-1}\mathfrak{d}\mathfrak{f}, \\ &= C(\chi; \delta, \varphi) h \theta(\alpha^{-1}\delta\varphi; \overline{\chi}).\end{aligned}$$

Now, as $\mathfrak{a} = \iota(\alpha) = \iota(\alpha_0)$, we have $V(\alpha_0) = N(\mathfrak{a})^{-1}$ for $\alpha = \alpha_0\alpha_\infty$. Also $\prod_{k=1}^n |t_k| = \prod_{k=1}^{r_1} |\sigma_k(a_{P_{\infty,k}})| \cdot \prod_{k=r_1+1}^{r_1+r_2} |\sigma_k(a_{P_{\infty,k}})|^2 = \prod_{k=1}^{r} v_{P_{\infty,k}}(a_{P_{\infty,k}}) = V(\alpha_\infty)$. Hence

$$\begin{aligned}c_1 \cdots c_n d N(\mathfrak{a}^{-1}\mathfrak{f})^2 &= \prod_{k=1}^{n} |t_k|^2 \cdot \frac{1}{df} \cdot d \cdot N(\mathfrak{a})^{-2} f^{-2} \\ &= V(\alpha_\infty)^2 \cdot V(\alpha_0)^2 \cdot f \\ &= V(\alpha)^2 f.\end{aligned}$$

On the other hand, since $\prod t_k^{m_k}$ is the complex conjugate of $\prod t_k^{m'_k}$, we have

$$\prod t_k^{m'_k} \left(\prod c_k^{m'_k}\right)^{-1} \prod \left(\frac{t_k}{\sqrt[n]{df}}\right)^{m_k} = \prod \left(\frac{|t_k|^2}{\sqrt[n]{df}}\right)^{m'_k} \left(\prod c_k^{m'_k}\right)^{-1} = 1.$$

Hence

$$h = \frac{(-i)^M}{V(\alpha)\sqrt{f}},$$

and the following result is proved.

Proposition 3.4 *Let α be any idèle of F, and let δ and φ be the fixed idèle as defined in Sect. 2.3. Then*

$$\theta(\alpha; \chi) = W'(\chi; \delta, \varphi) V(\alpha)^{-1} \theta(\alpha^{-1}\delta_0\varphi; \overline{\chi}),$$

where

$$W'(\chi;\delta,\varphi)=\frac{(-i)^M}{\sqrt{f}}C(\chi;\delta,\varphi),\quad M=\sum_{k=1}^{n}m_k,$$

depends only upon χ, δ, *and* φ, *and is independent of* α.

By Proposition 2.4, Sect. 2.3, we have

$$|C(\chi;\delta,\varphi)|=\sqrt{f},$$
$$\overline{C(\chi;\delta,\varphi)}=\chi_{\mathfrak{f}}(-1)C(\overline{\chi};\delta,\varphi).$$

Hence

$$|W'(\chi;\delta,\varphi)|=1,$$

and

$$\overline{W'(\chi;\delta,\varphi)}=\frac{i^M}{\sqrt{f}}\chi_{\mathfrak{f}}(-1)C(\overline{\chi};\delta,\varphi)=(-1)^M\chi_{\mathfrak{f}}(-1)W'(\overline{\chi};\delta,\varphi).$$

However, since $\chi_{\mathfrak{f}}(\xi)=\chi(\xi)$ for ξ in U_0, and $\chi(F^*)=1$, we see that

$$\begin{aligned}\chi_{\mathfrak{f}}(-1)&=\chi_{\mathfrak{f}}((-1)_0)=\chi((-1)_0)\\&=\chi((-1)_\infty)^{-1}=\left(\prod_{k=1}^{n}(-1)^{m_k}\right)^{-1}=((-1)^M)^{-1}.\end{aligned}$$

Therefore

$$\overline{W'(\chi;\delta,\varphi)}=W'(\overline{\chi};\delta,\varphi).$$

3.4 The Function $f(\alpha, s; x)$

Let F be a number field, $n=[F:\mathbf{Q}]$, and let χ be a Hecke character of F with conductor $\mathfrak{f}$ and signature $\{m_1,\ldots,m_n;s_1,\ldots,s_n\}$. The complex conjugate $\overline{\chi}$ of χ is then also a Hecke character of F, and its conductor is $\mathfrak{f}$ and its signature is $\{m'_1,\ldots,m'_n;-s_1,\ldots,-s_n\}$, when $m'_k=m_k$, $1\le k\le r_1$, $m'_k=m_{k+r_2}$, $m'_{k+r_2}=m_k$, $r_1+1\le k\le r_1+r_2$.

For any idèle α of F ($\alpha\in J$), and for any complex number s, define

$$f(\alpha,s;\chi)=\begin{cases}\chi(\alpha)g(\alpha;\chi)V(\alpha)^s, & \text{if } \mathfrak{a}=\iota(\alpha)\subset\mathfrak{o},\\ 0, & \text{otherwise.}\end{cases}$$

Here $V(\alpha)$ is the volume of α, and $g(\alpha;\chi)$ is the function defined in Sect. 3.3. Namely, let $\alpha=(a_P)$, $t_k=\sigma_k(a_{P_{\infty,k}})$, $1\le k\le r$, $t_{k+r_2}=\overline{t}_k$, $r_1+1\le k\le r$. Then

$$g(\alpha;\chi)=\overline{\chi}_{\mathfrak{f}}(\alpha)\prod_{k=1}^{n}t_k^{m'_k}\exp\left(-\frac{\pi}{\sqrt[n]{df}}\sum_{k=1}^{n}|t_k|^2\right).$$

Let $\alpha=\alpha_0\alpha_\infty$, $\alpha_0\in J_0$, $\alpha_\infty\in J_\infty$. Then $\chi(\alpha)=\chi(\alpha_0)\chi(\alpha_\infty)$, $\chi_{\mathfrak{f}}(\alpha)=\chi_{\mathfrak{f}}(\alpha_0)$, $V(\alpha)=V(\alpha_0)V(\alpha_\infty)$, and $\mathfrak{a}=\iota(\alpha)=\iota(\alpha_0)$. Hence

$$f(\alpha,s;\chi)=f_0(\alpha_0,s;\chi)f_\infty(\alpha_\infty,s;\chi),$$

where

$$f_0(\alpha,s;\chi)=\begin{cases}\chi(\alpha_0)\overline{\chi}_{\mathfrak{f}}(\alpha_0)V(\alpha_0)^s, & \text{if } \mathfrak{a}=\iota(\alpha_0)\subset\mathfrak{o},\\ 0, & \text{otherwise,}\end{cases}$$

$$f_\infty(\alpha_\infty,s;\chi)=\chi(\alpha_\infty)\prod t_k^{m'_k}\exp\left(-\frac{\pi}{\sqrt[n]{df}}|t_k|^2\right)V(\alpha_\infty)^s.$$

We first consider $f_0(\alpha_0)=f_0(\alpha_0,s;\chi)$. Suppose that $\mathfrak{a}=\iota(\alpha_0)\subset\mathfrak{o}$. Let $P_1,\dots,P_g$ be all the finite primes such that $\nu_{P_j}(\mathfrak{f})>0$. Since $\mathfrak{a}=\prod\mathfrak{p}_P^{\nu_P(a_P)}$, $\mathfrak{a}$ is prime to $\mathfrak{f}$ if and only if $\nu_{P_j}(a_{P_j})=0$ for $j=1,\dots,g$, namely, if and only if $(\alpha_0)_{\mathfrak{f}}=(a_{P_1},\dots,a_{P_g},1\dots,1,\dots)$ is in U_0. In such a case, $\chi_{\mathfrak{f}}(\alpha_0)=\chi((\alpha_0)_{\mathfrak{f}})$ so that $\chi(\alpha_0)\overline{\chi}_{\mathfrak{f}}(\alpha_0)=\chi(\alpha_0/(\alpha_0)_{\mathfrak{f}})$. However, $\alpha_0/(\alpha_0)_{\mathfrak{f}}=(1,\dots,1,a_P,\dots)$ is in $J_0(\mathfrak{f})$, and $\iota((\alpha_0)/(\alpha_0)_{\mathfrak{f}})=\iota(\alpha_0)\iota((\alpha_0)_{\mathfrak{f}})^{-1}=\iota(\alpha_0)=\mathfrak{a}$. Hence, by the definition of $\chi: J_{\mathfrak{f}}\to\mathbf{C}_1^*$, $\chi(\alpha_0/(\alpha_0)_{\mathfrak{f}})=\chi(\mathfrak{a})$, namely, $\chi(\alpha_0)\overline{\chi}_{\mathfrak{f}}(\alpha_0)=\chi(\mathfrak{a})$. On the other hand, if $\mathfrak{a}$ is not prime to $\mathfrak{f}$, then $\chi_{\mathfrak{f}}(\alpha_0)=0$ and $\chi(\mathfrak{a})=0$ by the definition of $\chi_{\mathfrak{f}}(\alpha_0)$ and $\chi(\mathfrak{a})$. Hence we still have $\chi(\alpha_0)\overline{\chi}_{\mathfrak{f}}(\alpha_0)=\chi(\mathfrak{a})$. Since $V(\alpha_0)=N(\mathfrak{a})^{-1}$, we obtain

$$f_0(\alpha_0)=\begin{cases}\chi(\mathfrak{a})N(\mathfrak{a})^{-s}, & \text{if } \mathfrak{a}=\iota(\alpha_0)\subset\mathfrak{o},\\ 0 & \text{otherwise.}\end{cases}$$

The homomorphism $\iota: J_0\to\mathfrak{J}$ induces an isomorphism $J_0/U_0\cong\mathfrak{J}$. The above equalities shows that $f_0(\alpha_0)$ depends only upon the coset $\overline{\alpha}_0=\alpha_0U_0$ in J_0/U_0; $f_0(\alpha_0)=f_0(\overline{\alpha}_0)$. Since U_0 is open in J_0, we see in particular that $f_0(\alpha_0,s;\chi)$ is a continuous function of (α_0,s) in $J_0\times\mathbf{C}$.

Let μ_{J_0}, μ_{U_0}, and μ_{J_0/U_0} be the Haar measures of the respective groups fixed in Sect. 2.4. Since $\mu_{J_0}=\mu_{J_0/U_0}*\mu_{U_0}$, we have

$$\int_{J_0}|f_0(\alpha_0)|d\mu_{J_0}(\alpha_0)=\int_{J_0/U_0}\left(\int_{U_0}|f_0(\alpha_0\xi)|d\mu_{U_0}(\xi)\right)d\mu_{J_0/U_0}(\overline{\alpha}_0).$$

However, $f_0(\alpha_0\xi)=f_0(\alpha_0)$ for any ξ in U_0, and $\mu_{U_0}(U_0)=1$. Hence the inner integral on the right is $|f_0(\alpha_0)|$. As μ_{J_0/U_0} is the point measure on the discrete group J_0/U_0, we see that

$$\int_{J_0} |f_0(\alpha_0, s; \chi)| d\mu_{J_0}(\alpha_0) = \int_{J_0/U_0} |f_0(\alpha_0)| d\mu_{J_0/U_0}(\overline{\alpha}_0)$$
$$= \sum_{\overline{\alpha}_0 \in J_0/U_0} |f_0(\overline{\alpha}_0)|$$
$$= \sum_{\mathfrak{a} \subset \mathfrak{o}} |\chi(\mathfrak{a}) N(\mathfrak{a})^{-s}|.$$

It follows from Sect. 3.1 that $\int_{J_0} |f_0(\alpha_0, s; \chi)| d\mu_{J_0}(\alpha_0) < +\infty$ if $\mathrm{Re}\,(s) > 1$. We also see from the above that in such a case,

$$\int_{J_0} f_0(\alpha_0, s; \chi) d\mu_{J_0}(\alpha_0) = \sum_{\mathfrak{a} \subset \mathfrak{o}} \chi(\mathfrak{a}) N(\mathfrak{a})^{-s}$$
$$= L(s; \chi).$$

We shall next consider $f_\infty(\alpha_\infty) = f_\infty(\alpha_\infty, s; \chi)$. By the definition of the signature, we have

$$\chi(\alpha_\infty) = \prod_{k=1}^{n} \left(\frac{t_k}{|t_k|}\right)^{m_k} |t_k|^{is_k}.$$

We also know that $V(\alpha_\infty) = \prod_{k=1}^{n} |t_k|$. Hence

$$f_\infty(\alpha_\infty) = \prod_{k=1}^{n} \left(\frac{t_k}{|t_k|}\right)^{m_k} |t_k|^{is_k} t_k^{m'_k} \exp\left(-\frac{\pi}{\sqrt[n]{df}}|t_k|^2\right) |t_k|^s$$
$$= \prod_{k=1}^{n} |t_k|^{s+m'_k+is_k} \exp\left(-\frac{\pi}{\sqrt[n]{df}}|t_k|^2\right)$$
$$= \prod_{k=1}^{r} |t_k|^{s+m'_k+is_k} \exp\left(-\frac{\pi}{\sqrt[n]{df}}|t_k|^2\right)$$
$$\times \prod_{k=r_1+1}^{r} |t_k|^{2s+m_k+m'_k+2is_k} \exp\left(-\frac{2\pi}{\sqrt[n]{df}}|t_k|^2\right).$$

This shows in particular that $f_\infty(\alpha_\infty, s; \chi)$ is a continuous function of (α_∞, s) in $J_\infty \times \mathbf{C}$. Hence $f(\alpha) = f_0(\alpha_0) f_\infty(\alpha_\infty)$ is also continuous on $J = J_0 \times J_\infty$.

Let μ_{J_∞} be the Haar measure of J_∞ fixed in Sect. 2.4. It follows from the definition of μ_{J_∞} that

$$\int_{J_\infty} f_\infty(\alpha_\infty) d\mu_{J_\infty}(\alpha_\infty) = I_1 I_2 \cdots I_r,$$

where

$$I_k = \int_{\mathbf{R}^*} |t|^{s+m'_k+is_k} \exp\left(-\frac{\pi}{\sqrt[n]{df}}|t|^2\right) d\mu(t), \qquad 1 \le k \le r_1,$$
$$= \int_{\mathbf{C}^*} |t|^{2s+m_k+m'_k+2is_k} \exp\left(-\frac{2\pi}{\sqrt[n]{df}}|t|^2\right) d\mu(t), \qquad r_1+1 \le k \le r,$$

μ being the standard measures on $\mathbf{R}^*$ and $\mathbf{C}^*$ respectively.

Let $1 \le k \le r_1$. Then $d\mu(t) = \frac{dt}{|t|}$. Hence

$$I_k = \int_{-\infty}^{0} + \int_{0}^{+\infty} = 2\int_0^\infty t^{s+m'_k+is_k} \exp\left(-\frac{\pi}{\sqrt[n]{df}}t^2\right)\frac{dt}{t}$$
$$= \left(-\frac{\sqrt[n]{df}}{\pi}\right)^{\frac{1}{2}(s+m'_k+is_k)} \int_0^\infty u^{\frac{1}{2}(s+m'_k+is_k)} \exp(-u)\frac{du}{u},$$

with $u = -\frac{\pi}{\sqrt[n]{df}}t^2$. Since $\mathrm{Re}\,(s+m'_k+is_k) = \mathrm{Re}\,(s)+m'_k \ge \mathrm{Re}\,(s)$, we see that if $\mathrm{Re}\,(s) > 0$, then the last integral is absolutely convergent, and

$$I_k = \left(\frac{\sqrt[n]{df}}{\pi}\right)^{\frac{1}{2}(s+m'_k+is_k)} \Gamma\left(\frac{1}{2}(s+m'_k+is_k)\right).$$

If $r_1+1 \le k \le r$, then $d\mu(t) = \frac{2dr d\theta}{r}$, where $t = re^{i\theta}$, $r = |t|$. Hence by a similar computation, we see that if $\mathrm{Re}\,(s) > 0$, then the integral for I_k converges absolutely, and

$$I_k = 2\pi\left(\frac{\sqrt[n]{df}}{2\pi}\right)^{s+\frac{m_k+m'_k}{2}+is_k} \Gamma\left(s+\frac{m_k+m'_k}{2}+is_k\right).$$

Thus we see that

$$\int_{J_\infty} |f_\infty(\alpha_\infty, s; \chi)| d\mu_{J_\infty}(\alpha_\infty) < +\infty$$

for $\mathrm{Re}\,(s) > 0$, and

$$\int_{J_\infty} f_\infty(\alpha_\infty, s; \chi) d\mu_{J_\infty}(\alpha_\infty) = I_1 \cdots I_r$$
$$= (2\pi)^{r_2} \prod_{k=1}^{r} \left(\frac{\sqrt[n]{df}}{e_k\pi}\right)^{\frac{e_k}{2}\left(s+\frac{m_k+m'_k}{2}+is_k\right)} \Gamma\left(-\frac{e_k}{2}\left(s+\frac{m_k+m'_k}{2}+is_k\right)\right),$$

where $e_k = 1$ or 2 according as $k \le r_1$ or $k > r_1$. We may also write it in the form

$$\int_{J_\infty} f_\infty(\alpha_\infty, s; \chi) d\mu_{J_\infty}(\alpha_\infty) = A(s; \chi)\gamma(s; \chi),$$

where

$$A(s; \chi) = (2\pi)^{r_2} \prod_{k=1}^{r} \left(\frac{\sqrt[n]{df}}{e_k \pi} \right)^{\frac{e_k}{2}\left(s + \frac{m_k + m_k'}{2} + i s_k\right)},$$

$$\gamma(s; \chi) = \prod_{k=1}^{r} \Gamma\left(\frac{e_k}{2} \left(s + \frac{m_k + m_k'}{2} + i s_k \right) \right).$$

For example, if $\chi \equiv 1$, then $m_k = m_k' = s_k = 0$, $1 \le k \le r$, so that

$$A(s; 1) = (2\pi)^{r_2} \left(\frac{\sqrt{d}}{2^{r_2} \pi^{\frac{n}{2}}} \right)^s,$$

$$\gamma(s; 1) = \Gamma\left(\frac{s}{2}\right)^{r_1} \Gamma(s)^{r_2}.$$

Now, since $f(\alpha, s; \chi) = f_0(\alpha_0, s; \chi) f_\infty(\alpha_\infty, s; \chi)$, $J = J_0 \times J_\infty$, $\mu_J = \mu_{J_0} \times \mu_{J_\infty}$, we immediately obtain from the above the following result.

Proposition 3.5 *The function $f(\alpha, s; \chi)$ is continuous on $J \times \mathbf{C}$. For* Re $(s) > 1$, *it is integrable on J:*

$$\int_J |f(\alpha, s; \chi)| d\mu_J(\alpha) < +\infty, \quad \mathrm{Re}\,(s) > 1.$$

Let

$$\xi(s; \chi) = \int_J f(\alpha, s; \chi) d\mu_J(\alpha), \quad \mathrm{Re}\,(s) > 1.$$

Then

$$\xi(s; \chi) = A(s; \chi)\gamma(s; \chi)L(s; \chi), \quad \mathrm{Re}\,(s) > 1,$$

where $A(s; \chi)$ and $\gamma(s; \chi)$ are the functions defined in the above.

Let

$$O = \{\alpha \mid \alpha \in J, V(\alpha) > 1\}.$$

O is an open set in J.

Proposition 3.6 *For any complex s, $f(\alpha, s; \chi)$ is integrable on O, and*

$$\eta(s; \chi) = \int_O f(\alpha, s; \chi) d\mu(\alpha)$$

defines a holomorphic function of s on the entire s-plane.

Proof Given any $s \in \mathbf{C}$, find $c > 1$ such that $\mathrm{Re}(s) \le c$. As

$$|f(\alpha, s; \chi)| = \begin{cases} |\chi(\alpha) g(\alpha; \chi) V(\alpha)^s| = |g(\alpha; \chi)| V(\alpha)^{\mathrm{Re}(s)}, & \text{or} \\ 0, & \end{cases}$$

we see that

$$|f(\alpha, s; \chi)| \le |f(\alpha, c; \chi)|$$

for α in O; $V(\alpha) > 1$. Hence

$$\begin{aligned} \int_O |f(\alpha, s; \chi)| d\mu_J(\alpha) &\le \int_O |f(\alpha, c; \chi)| d\mu_J(\alpha) \\ &\le \int_J |f(\alpha, c; \chi)| d\mu_J(\alpha) < +\infty, \end{aligned}$$

by Proposition 3.5. This proves the first half. At the same time, it was shown that

$$|\eta(s; \chi)| \le \int_O |f(\alpha, s; \chi)| d\mu_J(\alpha) \le K_c < +\infty$$

whenever $\mathrm{Re}(s) \le c$.

Let C be an arbitrary circle on the s-plane, and let $c > 1$ be chosen so that $\mathrm{Re}(s) \le c$ for $s \in C$. For fixed $\alpha \in O$, $f(\alpha, s; \chi)$ is of course holomorphic in s. Hence, for any s inside of the circle C, we have

$$f(\alpha, s; \chi) = \frac{1}{2\pi i} \int_C \frac{f(\alpha, z; \chi)}{z - s} dz.$$

Therefore

$$\eta(s; \chi) = \frac{1}{2\pi i} \int_O \left(\int_C \frac{f(\alpha, z; \chi)}{z - s} dz \right) d\mu_J(\alpha).$$

However,

$$\int_{O \times C} \left| \frac{f(\alpha, z; \chi)}{z - s} \right| d\mu_J(\alpha) dz \le \int_C \frac{K_c}{|z - s|} dz < +\infty.$$

Hence, by Fubini's Theorem,

$$\begin{aligned} \eta(s; \chi) &= \frac{1}{2\pi i} \int_C \frac{1}{z - s} \left(\int_O f(\alpha, z; \chi) d\mu_J(\alpha) \right) dz \\ &= \frac{1}{2\pi i} \int_C \frac{\eta(z; \chi)}{z - s} dz. \end{aligned}$$

This holds for any s inside of C. As $|\eta(z; \chi)| \leq K_c$ on C, we see that $\eta(s; \chi)$ is holomorphic in s inside of C, and hence it is holomorphic everywhere on the s-plane.

Let μ_{F^*} denote as before the point measure of the discrete subgroup F^* of J, and let $\mu_{\overline{J}} = \mu_J/\mu_{F^*}$, $\overline{J} = J/F^*$. Since $V(\alpha a) = V(\alpha)$, $\alpha \in J$, $a \in F^*$, O is a union of cosets mod F^*. Hence

$$\int_O \cdots d\mu_J = \int_{\overline{O}} \left(\int_{F^*} \cdots d\mu_{F^*} \right) d\mu_{\overline{J}},$$

where $\overline{O}$ denotes the image of O under $J \to \overline{J} = J/F^*$. In particular,

$$\eta(s; \chi) = \int_{\overline{O}} \left(\int_{F^*} f(\alpha a, s; \chi) d\mu_{F^*}(a) \right) d\mu_{\overline{J}}(\overline{\alpha}).$$

Since μ_{F^*} is the point measure,

$$\begin{aligned} \int_{F^*} f(\alpha a, s; \chi) d\mu_{F^*}(a) &= \sum_{a \in F^*} f(\alpha a, s; \chi) \\ &= \sum_{\substack{a \in \mathfrak{a}^{-1} \\ a \neq 0}} \chi(\alpha a) g(\alpha a; \chi) V(\alpha a)^s \\ &= \chi(\alpha) V(\alpha)^s \sum_{\substack{a \in \mathfrak{a}^{-1} \\ a \neq 0}} g(\alpha a; \chi) \\ &= \chi(\alpha) V(\alpha)^s (\theta(\alpha; \chi) - \varepsilon_\chi), \end{aligned}$$

where $\theta(\alpha; \chi) = \sum_{a \in \mathfrak{a}^{-1}} g(\alpha a; \chi)$ as in Sect. 3.3, and $\varepsilon_\chi = g(0; \chi)$. Hence we have the following formula:

$$\eta(s; \chi) = \int_{\overline{O}} \chi(\alpha)(\theta(\alpha; \chi) - \varepsilon_\chi) V(\alpha)^s d\mu_{\overline{J}}(\overline{\alpha}).$$

Note that the functions $\chi(\alpha)$, $V(\alpha)$, and $\theta(\alpha; \chi)$ depend only upon the coset $\overline{\alpha} = \alpha F^*$ of α mod F^* so that they may be also denoted by $\chi(\overline{\alpha})$, $V(\overline{\alpha})$, and $\theta(\overline{\alpha}; \chi)$. Note also that since

$$g(\alpha; \chi) = \overline{\chi}_{\mathfrak{f}}(\alpha) \prod_{k=1}^{n} t_k^{m'_k} \exp\left(-\frac{\pi}{\sqrt[n]{df}} |t_k|^2 \right),$$

we have

$$\varepsilon_\chi = g(0; \chi) = \begin{cases} 1, & \text{if } \mathfrak{f} = \mathfrak{o} \text{ and } m_k = 0, \quad 1 \leq k \leq n, \\ 0, & \text{otherwise.} \end{cases}$$

3.5 Fundamental Theorems

Now, we have the disjoint decomposition of J; $J = O \cup O^{-1} \cup J_1$, so that

$$\int_J = \int_O + \int_{O^{-1}} + \int_{J_1}.$$

However, since $J = J_1 \times T$, $J_1 = J_1 \times \{1\}$, $\mu_J = \mu_{J_1} \times \mu_T$, and since the point 1 has measure 0 in T, we see that $\mu_J(J_1) = 0$, $\int_{J_1} = 0$. Hence, in particular,

$$\begin{aligned}\xi(s;\chi) &= \int_J f(\alpha, s; \chi) d\mu_J(\alpha)\\ &= \int_O f(\alpha, s; \chi) d\mu_J(\alpha) + \int_{O^{-1}} f(\alpha, s; \chi) d\mu_J(\alpha),\end{aligned}$$

for any s with $\mathrm{Re}\,(s) > 1$. The integral over O is equal to $\eta(s;\chi)$. For the integral over O^{-1}, we obtain

$$\int_{O^{-1}} f(\alpha, s; \chi) d\mu_J(\alpha) = \int_{\overline{O}^{-1}} \chi(\alpha)(\theta(\alpha;\chi) - \varepsilon_\chi) V(\alpha)^s d\mu_{\overline{J}}(\overline{\alpha}),$$

in the same way as we have done in Sect. 3.4. However,

$$\theta(\alpha;\chi) = W'(\chi;\delta,\varphi) V(\alpha)^{-1} \theta(\alpha^{-1}\delta\varphi; \overline{\chi})$$

by Proposition 3.4, Sect. 3.3. Hence

$$\int_{O^{-1}} f(\alpha, s; \chi) d\mu_J(\alpha) = \int_{\overline{O}^{-1}} \chi(\alpha)(W' V(\alpha)^{-1} \theta(\alpha^{-1}\delta\varphi; \overline{\chi}) - \varepsilon_\chi) V(\alpha)^s d\mu_{\overline{J}}(\overline{\alpha}).$$

It follows from the last remark in Sect. 3.4 that $\varepsilon_\chi = \varepsilon_{\overline{\chi}} = 1$ or 0 and that if $\varepsilon_\chi = 1$, then $\mathfrak{f} = \mathfrak{o}$, $m_k = 0$, $1 \le k \le n$, so that

$$W'(\chi;\delta,\varphi) = -\frac{(-i)^{\sum m_k}}{\sqrt{f}} C(\chi;\delta,\varphi) = 1.$$

Hence we may write the integrand of the last integral in the form

$$W'\chi(\alpha)(\theta(\alpha^{-1}\delta\varphi;\overline{\chi}) - \varepsilon_{\overline{\chi}}) V(\alpha)^{s-1} + \varepsilon_\chi \chi(\alpha)(V(\alpha)^{s-1} - V(\alpha)^s).$$

Replace α by $\alpha^{-1}\delta\varphi$ in the first term of the above. Since $\delta, \varphi \in J_1$, $V(\delta\varphi) = 1$, we then obtain

$$W'\chi(\delta\varphi)\overline{\chi}(\alpha)(\theta(\alpha;\overline{\chi}) - \varepsilon_{\overline{\chi}}) V(\alpha)^{1-s}.$$

As $O^{-1} \to O$ ($\overline{O}^{-1} \to \overline{O}$) under $\alpha \mapsto \alpha^{-1}\delta\varphi$ ($\overline{\alpha} \mapsto \overline{\alpha}^{-1}\overline{\delta\varphi}$), we see that

$$\begin{aligned}
&\int_{\overline{O}^{-1}} W'\chi(\alpha)(\theta(\alpha^{-1}\delta\varphi;\overline{\chi}) - \varepsilon_{\overline{\chi}})V(\alpha)^{s-1}d\mu_{\overline{J}}(\overline{\alpha}) \\
&\quad = W\int_{\overline{O}} \overline{\chi}(\alpha)(\theta(\alpha;\overline{\chi}) - \varepsilon_{\overline{\chi}})V(\alpha)^{1-s}d\mu_{\overline{J}}(\overline{\alpha}) \\
&\quad = W\eta(1-s;\overline{\chi}),
\end{aligned}$$

with

$$W = W(\chi) = \chi(\delta\varphi)W'(\chi;\delta\varphi).$$

Problem 3.1 *Prove that $W(\chi)$ depends only upon χ, and is independent of the choice of δ and φ such that $\delta, \varphi \in J_1 \cap (J_0 \times T)$, $\iota(\delta) = \mathfrak{d}$, $\iota(\varphi) = \mathfrak{f}$.*

It follows that

$$\begin{aligned}
\xi(s;\chi) &= \int_{O} + \int_{O^{-1}} \\
&= \eta(s;\chi) + W\eta(1-s;\overline{\chi}) + \int_{O^{-1}} \varepsilon_\chi \chi(\alpha)\left(V(\alpha)^{s-1} - V(\alpha)^s\right)d\mu_{\overline{J}}(\overline{\alpha}),
\end{aligned}$$

for any s with Re $(s) > 1$. The computation also shows that the last integral converges for Re $(s) > 1$.

Now, let

$$\overline{\alpha} = \overline{\alpha}_1 \cdot \overline{\tau}_x, \quad \overline{\alpha}_1 \in \overline{J}_1, \overline{\tau}_x \in \overline{T}, x \in \mathbf{R}^+,$$

for $\overline{\alpha}$ in $\overline{J} = \overline{J}_1 \times \overline{T}$. Then $V(\overline{\alpha}) = V(\overline{\tau}_x) = x$, and we have

$$\begin{aligned}
\overline{O}^{-1} &= \{\overline{\alpha} \mid \overline{\alpha} \in \overline{J}, V(\overline{\alpha}) < 1\} \\
&= \overline{J}_1 \times \{\overline{\tau}_x \mid 0 < x < 1\}.
\end{aligned}$$

Since $d\mu_{\overline{T}}(\overline{\tau}_x) \mapsto \frac{dx}{x}$ under $\overline{T} \cong \mathbf{R}^+$, we see that

$$\begin{aligned}
&\int_{\overline{O}^{-1}} \varepsilon_\chi \chi(\alpha)\left(V(\alpha)^{s-1} - V(\alpha)^s\right)d\mu_{\overline{J}}(\overline{\alpha}) \\
&\quad = \int_{\overline{J}_1} \varepsilon_\chi \chi(\overline{\alpha}_1)d\mu_{\overline{J}_1}(\overline{\alpha}_1)\int_0^1 \chi(\overline{\tau}_x)(x^{s-1} - x^s)\frac{dx}{x}, \qquad (\chi(\overline{\tau}_x) = 1).
\end{aligned}$$

Suppose first that $\chi \equiv 1$. Then $\varepsilon_\chi = 1$, and the above integrals become

$$\begin{aligned}
\int_{\overline{J}_1} d\mu_{\overline{J}_1}(\overline{\alpha}_1) \int_0^1 (x^{s-1} - x^s)\frac{dx}{x} \qquad (\because \mathrm{Re}\,(s) > 1) \\
&= \mu_{\overline{J}_1}(\overline{J}_1) \cdot \left[\frac{x^{s-1}}{s-1} - \frac{x^s}{s}\right]_0^1 \\
&= \mu_{\overline{J}_1}(\overline{J}_1)\left(\frac{1}{s-1} - \frac{1}{s}\right) \\
&= \frac{1}{s(s-1)}\mu_{\overline{J}_1}(\overline{J}_1).
\end{aligned}$$

Since this integral must have a finite value, we see that

$$\mu_{\overline{J}_1}(\overline{J}_1) < +\infty.$$

Suppose next that $\chi \not\equiv 1$. As χ is a Hecke character, $\chi(F^* \times T) = 1$. Hence χ must induce a non-trivial character on $\overline{J}_1 = J_1/F^*$. Let $\overline{\beta}_1$ be an element of $\overline{J}_1$ such that $\chi(\overline{\beta}_1) \neq 1$. Since $\mu_{\overline{J}_1}(\overline{J}_1) < +\infty$, the bounded continuous function $\chi(\overline{\alpha}_1)$, $\overline{\alpha}_1 \in \overline{J}_1$, is integrable on $\overline{J}_1$, and

$$\begin{aligned}
\int_{\overline{J}_1} \chi(\overline{\alpha}_1) d\mu_{\overline{J}_1}(\overline{\alpha}_1) &= \int_{\overline{J}_1} \chi(\overline{\alpha}_1\overline{\beta}_1) d\mu_{\overline{J}_1}(\overline{\alpha}_1) \\
&= \chi(\overline{\beta}_1) \int_{\overline{J}_1} \chi(\overline{\alpha}_1) d\mu_{\overline{J}_1}(\overline{\alpha}_1).
\end{aligned}$$

Then it follows from $\chi(\overline{\beta}_1) \neq 1$ that

$$\int_{\overline{J}_1} \chi(\overline{\alpha}_1) d\mu_{\overline{J}_1}(\overline{\alpha}_1) = 0,$$

and that $\int_{\overline{J}_1} \varepsilon_\chi \chi(\overline{\alpha}_1) d\mu_{\overline{J}_1}(\overline{\alpha}_1) = \varepsilon_\chi \int_{\overline{J}_1} \chi(\overline{\alpha}_1) d\mu_{\overline{J}_1}(\overline{\alpha}_1) = 0.$

Thus we see that

$$\xi(s;\chi) = \eta(s;\chi) + W(\chi)\eta(1-s;\overline{\chi}) + \frac{\varepsilon_\chi v}{s(s-1)}, \quad \mathrm{Re}\,(s) > 1. \tag{3.5}$$

where $v = \mu_{\overline{J}_1}(\overline{J}_1) < +\infty$ and $\varepsilon_\chi = 1$ or 0 according as $\chi \equiv 1$ or $\chi \not\equiv 1$. By Proposition 3.6, the right-hand side of the above is a meromorphic function of s on the entire s-plane; if $\chi \equiv 1$, it has only simple poles at $s = 0$ and $s = 1$, and if $\chi \not\equiv 1$, it is holomorphic everywhere. Hence we see from (3.5) that the function $\xi(s;\chi)$, which was originally defined only for complex s with $\mathrm{Re}\,(s) > 1$ analytically continued to a meromorphic function of s on the entire s-plane, and satisfies (3.5) for arbitrary s.

Now, $W(\chi) = \chi(\delta\varphi)W'(\chi;\delta,\varphi)$ by the definition. Hence it follows from Proposition 2.4, that

$$|W(\chi)| = |\chi(\delta\varphi)||W'(\chi; \delta, \varphi)| = 1,$$
$$\overline{W(\chi)} = \overline{\chi}(\delta\varphi)W'(\overline{\chi}; \delta, \varphi) = W(\overline{\chi}),$$

so that

$$W(\chi)W(\overline{\chi}) = 1.$$

On the other hand, if $\chi \equiv 1$, then $W(1) = 1$. Hence we obtain from (3.5) that

$$\begin{aligned}\xi(s; \chi) &= W(\chi)\left(W(\overline{\chi})\eta(s; \chi) + \eta(1-s; \overline{\chi}) + \frac{\varepsilon_\chi \nu}{s(s-1)}\right)\\ &= W(\chi)\xi(1-s; \overline{\chi}),\end{aligned}$$

namely,

$$\xi(s; \chi) = W(\chi)\xi(1-s; \overline{\chi}).$$

By Proposition 3.5, we have

$$L(s; \chi) = A(s; \chi)^{-1}\gamma(s; \chi)^{-1}\xi(s; \chi). \tag{3.6}$$

for Re $(s) > 1$. Here $A(s; \chi)^{-1}$ is a function of the form e^{as+b}, where $a, b =$ constant, and $\gamma(s; \chi)^{-1}$ is a product of functions of the form $\Gamma(cs+d)^{-1}$. Since $\Gamma(s)^{-1}$ is holomorphic on the entire s-plane, $A(s; \chi)^{-1}\gamma(c; \chi)^{-1}$ is also holomorphic for arbitrary s. It then follows from the above that the L-function $L(s; \chi)$, which was originally defined for s with Re $(s) > 1$, is a meromorphic function of s on the entire s-plane, satisfying (3.6) for arbitrary s. If $\chi \not\equiv 1$, then $\xi(s; \chi)$ is an entire function of s. Hence $L(s; \chi)$ is also an entire function of s. Let $\chi \equiv 1$. $\zeta_F(s) = L(s; 1)$ is holomorphic at $s \neq 0, 1$, and since

$$\gamma(s; 1)^{-1} = \Gamma\left(\frac{s}{2}\right)^{-r_1}\Gamma(s)^{-r_2}, \quad r_1 + r_2 > 0,$$

has a zero at $s = 0$, $\zeta_F(s)$ is still holomorphic at $s = 0$. At $s = 1$, $\xi(s; 1)$ has a simple pole with residue $\nu = \mu_{\overline{J}_1}(\overline{J}_1)$, and

$$A(1, 1)^{-1} = \frac{\pi^{\frac{r_1}{2}}}{\sqrt{d}}, \quad \gamma(1; 1)^{-1} = \Gamma\left(\frac{1}{2}\right)^{-r_1} = \pi^{-\frac{r_1}{2}}.$$

Hence $\zeta_F(s)$ has a simple pole at $s = 1$ with residue $\frac{\nu}{\sqrt{d}}$.

We now summarize our results as follows.

Theorem 3.1 *The L-function $L(s; \chi)$ for a Hecke character χ of F, which was originally defined for s with* Re $(s) > 1$, *is a meromorphic function of s on the entire s-plane. If $\chi \not\equiv 1$, then $L(s; \chi)$ is holomorphic everywhere (i.e., an entire function of s). If $\chi \equiv 1$, then $\zeta_F(s) = L(s; 1)$ has a unique simple pole at $s = 1$ with residue $\frac{\nu}{\sqrt{d}}$, $\nu = \mu_{\overline{J}_1}(\overline{J}_1) < +\infty$. Let*

$$\xi(s;\chi) = A(s;\chi)\gamma(s;\chi)L(s;\chi), \quad s \in \mathbf{C},$$

with $A(s;\chi)$ and $\gamma(s;\chi)$ in Sect. 3.3. Then $\xi(s;\chi)$ satisfies the functional equation

$$\xi(s;\chi) = W(\chi)\xi(1-s;\overline{\chi}), \quad s \in \mathbf{C},$$

where $W(\chi)$ is a constant, depending only upon χ, such that

$$|W(\chi)| = 1, \quad \overline{W(\chi)} = W(\overline{\chi}).$$

Now, since $m_k, m'_k \geq 0$ and s_k are real,

$$\gamma(s;\chi) = \prod_{k=1}^{r} \Gamma\left(\frac{e_k}{2}\left(s + \frac{m_k + m'_k}{2} + is_k\right)\right)$$

does not vanish for Re $(s) > 1$. Hence $\xi(s;\chi) = A(s;\chi)\gamma(s;\chi)L(s;\chi)$ also has no zero for Re $(s) > 1$. By the functional equation, we then see that $\xi(s;\chi)$ has no zero for Re $(s) < 0$. Hence the zeros of $L(s;\chi)$ in the domain Re $(s) < 0$ are obtained from those of $\gamma(s;\chi)^{-1}$, and they can be easily determined because $\Gamma(s)^{-1}$ has only simple zeros at $s = 0, -1, -2, \ldots$. Thus all "non-trivial" zeros of $L(s;\chi)$ are in the critical strip $\{s \mid 0 \leq \text{Re}\,(s) \leq 1\}$. The generalized Riemann hypothesis states that all such zeros are on the straight line Re $(s) = \frac{1}{2}$.

In general, a locally compact group is compact if and only if it has a finite total Haar measure. In the above, we have shown that $\mu_{\overline{J}_1}(\overline{J}_1) < +\infty$. Hence

Theorem 3.2 *The group $\overline{J}_1 = J_1/F^*$ is a compact group.*

We shall next show that the compactness of $\overline{J}_1$ implies two fundamental theorems on algebraic number fields, namely, the finiteness of class numbers and Dirichlet's unit theorem.

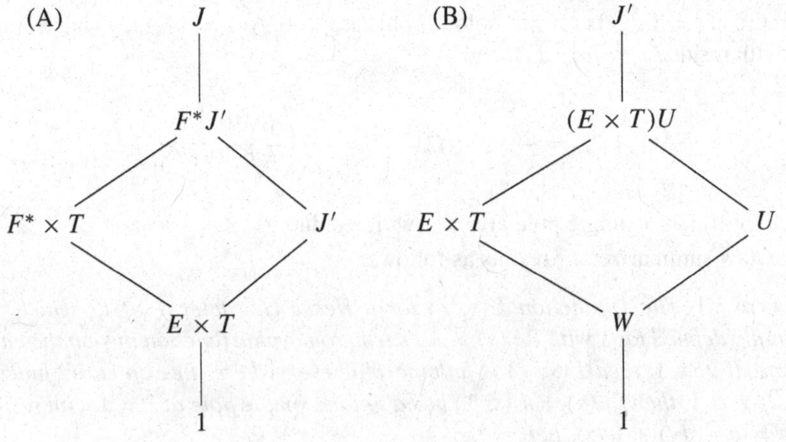

Let $J' = U_0 \times J_\infty$ and U be as before. Consider the diagrams in the above. Since $T \subset J'_\infty \subset J'$, we have $(F^* \times T)J' = F^*J'$, and $(F^* \times T) \cap J' = (F^* \cap J') \times T = E \times T$, where $E = F^* \cap J'$ is the group of units of F. Since $U \subset J_1$ and $J = J_1 \times T$, we have $(E \times T) \cap U = E \cap U = W$, which is the group of all roots of unity in F.

Since J' is open in J, F^*J' is an open (hence also closed) subgroup of J. Therefore J/F^*J' is discrete. Since U is compact, $(E \times T)U$ is closed, and U/W is compact. As J is separable, the two diagrams gives us topological isomorphisms

$$F^*J'/F^* \times T \cong J'/E \times T, \quad (E \times T)U/E \times T \cong U/W.$$

Hence $(E \times T)U/E \times T$ is compact.

Let

$$\lambda : J' \to \mathbf{R}^r, \quad r = r_1 + r_2,$$

be defined by

$$\alpha = (a_P) \mapsto (\log |t_1|, \ldots, \log |t_r|), \quad t_k = \sigma_k(a_{P_{\infty,k}}).$$

λ is a surjective continuous homomorphism with kernel U. Since J' is separable, λ induces a topological isomorphism

$$\lambda' : J'/U \to \mathbf{R}^r.$$

Let $H = \lambda(J_1 \cap J')$ and $L = \lambda(T)$. Then H is the hyperplane consisting of all vectors $(x_1, \ldots, x_r)$, $x_k \in \mathbf{R}$, satisfying $\sum_{k=1}^r e_k x_k = 0$, where $e_k = 1$ or 2 according as $1 \le k \le r_1$ or $r_1 + 1 \le k \le r$, and L is the straight line generated by $\omega = (\frac{1}{n}, \ldots, \frac{1}{n})$: $L = \mathbf{R}\omega$. Therefore

$$\mathbf{R}^r = H \oplus L.$$

Since $E = F^* \cap J' \subset J_1 \cap J'$, $\lambda(E)$ is contained in H. Hence

$$\lambda((E \times T)U) = \lambda(E \times T) = \lambda(E) \oplus L,$$

and λ' induces a topological isomorphism

$$J'/(E \times T)U \cong (H \oplus L)/(\lambda(E) \oplus L) = H/\lambda(E).$$

On the other hand, λ' also induces a topological isomorphism

$$E/W \cong EU/U \cong \lambda(E).$$

Since E is discrete, we see that $\lambda(E)$ is a discrete subgroup of $H \cong \mathbf{R}^{r-1}$.

Now, since $J = J_1 \times T$, we have

$$\overline{J}_1 = J_1/F^* \cong J/F^* \times T.$$

Hence

$$\begin{aligned}\text{compactness of } \overline{J}_1 &\Longleftrightarrow \text{compactness of } J/F^* \times T\\ &\Longleftrightarrow \text{compactness of } J/F^*J' \text{ and } F^*J'/F^* \times T\\ &\Longleftrightarrow \text{compactness of } J/F^*J' \text{ and } J'/E \times T.\end{aligned}$$

Since $(E \times T)U/E \times T$ is compact,

$$\begin{aligned}\text{compactness of } J'/E \times T &\Longleftrightarrow \text{compactness of } J'/(E \times T)U\\ &\Longleftrightarrow \text{compactness of } H/\lambda(E).\end{aligned}$$

Thus

$$\text{compactness of } \overline{J}_1 \Longleftrightarrow \text{compactness of } J/F^*J' \text{ and } H/\lambda(E).$$

Now, since J/F^*J' is discrete, the compactness of J/F^*J' implies that it is a finite group. However, we know from Chap. 2 that J/F^*J' is naturally isomorphic to the ideal class group $\mathfrak{I}/\mathfrak{H}$ of F. Therefore the class number $h = [\mathfrak{I} : \mathfrak{H}] = [J : F^*J']$ of F is finite. On the other hand, since $\lambda(E)$ is a discrete subgroup of $H \cong \mathbf{R}^{r-1}$, the compactness of $H/\lambda(E)$ implies that $\lambda(E)$ is an $(r-1)$-dimensional lattice in $H \cong \mathbf{R}^{r-1}$. Hence $\lambda(E)$ is a free abelian group of rank $r-1$, and consequently E/W has the same structure. This is Dirichlet's unit theorem.

Although the above proof may not be simpler than the usual proof of these classical results, it is quite interesting because it clearly indicates the remarkable fact that the two main theorems in algebraic number theory, the finiteness of class numbers and Dirichlet's unit theorem, are consequences of the compactness of the group $\overline{J}_1$, which in turn is a consequence of the finiteness of the integral $\int_J f(\alpha, s; 1) d\mu_J(\alpha)$ for $\mathrm{Re}\,(s) > 1$, namely, essentially a consequence of the convergence of $\zeta_F(s) = \sum N(\mathfrak{a})^{-s}$ for $\mathrm{Re}\,(s) > 1$.

3.6 The Residue of $\zeta_F(s)$ at $s = 1$

We shall next indicate how to compute $v = \mu_{\overline{J}_1}(\overline{J}_1)$ explicitly. By Theorem 3.1, this will give us an explicit value of the residue of $\zeta_F(s)$ at $s = 1$.

In general, let G be a separable (or σ-compact) locally compact abelian group, and let U and V be closed subgroups of G such that UV is also closed in G. Let μ_G, μ_U, μ_V and $\mu_{U \cap V}$ be arbitrary Haar measures on G, U, V, and $U \cap V$ respectively.

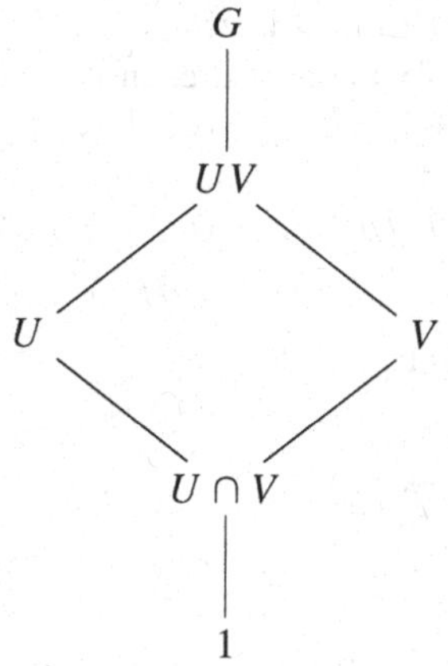

Lemma 3.3 *To each pair of groups X, Y, $Y \subset X$, in the diagram on the above, there exists a unique Haar measure $\mu_{X/Y}$ such that*

(i) *For $X = G, U, V, U \cap V$, $\mu_{X/1}$ coincides with the given μ_X,*
(ii) *For any $Z \subset Y \subset X$, $\mu_{X/Y} \cdot \mu_{Y/Z} = \mu_{X/Z}$,*
(iii) $\mu_{UV/U} \approx \mu_{V/U\cap V}$, $\mu_{UV/V} \approx \mu_{U/U\cap V}$ *in the obvious manner.*

Proof Let

$$\mu'_{U/U\cap V} = \mu_U/\mu_{U\cap V}, \quad \mu'_{V/U\cap V} = \mu_V/\mu_{U\cap V}, \quad \mu'_{UV/U\cap V} = \mu'_{U/U\cap V} \times \mu'_{V/U\cap V},$$

and let $\mu_{UV} = \mu'_{UV/U\cap V} \times \mu_{U\cap V}$. Let μ_1 be the point measure of 1: $\mu_1(1) = 1$. With these μ_G, μ_{UV}, μ_U, μ_V, $\mu_{U\cap V}$, μ_1, put

$$\mu_{X/Y} = \mu_X/\mu_Y,$$

for any X, Y, $Y \subset X$. Then (i) is trivially satisfied. Also $\mu_{X/Y} \cdot \mu_{Y/Z} = (\mu_X/\mu_Y) \cdot (\mu_Y/\mu_Z) = \mu_X/\mu_Z = \mu_{X/Z}$ (cf. Sect. 2.4). It is known that $\mu'_{UV/U\cap V}/\mu'_{U/U\cap V} \approx \mu'_{V/U\cap V}$. However, $\mu'_{U/U\cap V} = \mu_{U/U\cap V}$, $\mu'_{V/U\cap V} = \mu_{U/U\cap V}$, and $\mu'_{UV/U\cap V} = \mu_{UV}/\mu_{U\cap V} = \mu_{UV/U\cap V}$. Hence $\mu_{V/U\cap V} \approx \mu_{UV/U\cap V}/\mu_{U/U\cap V} = (\mu_{UV}/\mu_{U\cap V})/(\mu_U/\mu_{U\cap V}) = \mu_{UV}/\mu_U = \mu_{UV/U}$ (Sect. 2.4). Similarly, $\mu_{U/U\cap V} \approx \mu_{UV/V}$. The uniqueness is obvious, because there is no other choice for μ_{UV} and μ_1.

In Sect. 2.4, we have fixed the Haar measure μ_G for $G = U_0$, J_0/U_0, J_0, J_∞, T, J_1, F^*, $\overline{J}$, $\overline{J}_1$, and T. Let

$$\mu_{J'} = \mu_{U_0} \times \mu_{J_\infty}$$
$$\mu_{F^* \times T} = \mu_{F^*} \times \mu_T,$$
$$\mu_{E \times T} = \mu_E \times \mu_T,$$

where μ_E is the point measure of the discrete group E. We apply the lemma to the diagram (A) of Sect. 3.5, and fix the measures of the groups in that diagram by the above lemma. Since $J = J_1 \times T$, we have

$$\begin{aligned}\mu_{J/(F^*\times T)} &= \mu_J/\mu_{F^*\times T} = (\mu_{J_1} \times \mu_T)/(\mu_{F^*} \times \mu_T)\\ &\approx (\mu_{J_1}/\mu_T) \times (\mu_T/\mu_T)\\ &= \mu_{J_1}/\mu_T = \mu_{\overline{J}_1}.\end{aligned}$$

Hence

$$v = \mu_{\overline{J}_1}(\overline{J}_1) = \mu_{J/(F^*\times\mu_T)}(J/(F^* \times T)).$$

Also

$$\begin{aligned}\mu_{(F^*\times T)/(E\times T)} &= \mu_{F^*\times T}/\mu_{E\times T} = (\mu_{F^*} \times \mu_T)/(\mu_E \times \mu_T)\\ &\approx (\mu_{F^*}/\mu_E) \times (\mu_T/\mu_T)\\ &= \mu_{F^*}/\mu_E.\end{aligned}$$

Since μ_{F^*} and μ_E are both point measures, so is μ_{F^*}/μ_E. Hence $\mu_{(F^*\times T)/(E\times T)}$, and consequently $\mu_{F^*J'/J'}$ are also point measures. As

$$\begin{aligned}\mu_{J/J'} &= \mu_J/\mu_{J'} = (\mu_{J_0} \times \mu_{J_\infty})/(\mu_{U_0} \times \mu_{J_\infty})\\ &\approx \mu_{J_0}/\mu_{U_0}\\ &= \mu_{J_0/U_0},\end{aligned}$$

$\mu_{J/J'}$ is a point measure. Since $\mu_{J/J'} = \mu_{J/F^*J'} \cdot \mu_{F^*J'/J'}$, we see that $\mu_{J/F^*J'}$ is also a point measure. Hence

$$\begin{aligned}v &= \mu_{J/(F^*\times T)}(J/(F^* \times T))\\ &= \mu_{J/F^*J'}(J/F^*J') \cdot \mu_{F^*J'/(F^*\times T)}(F^*J'/(F^* \times T))\\ &= [J : F^*J']\mu_{J'/(E\times T)}(J'/(E \times T))\\ &= h\mu_{J'/(E\times T)}(J'/(E \times T)).\end{aligned}$$

Let J'_∞ be the inverse image of $(\mathbf{R}^+)^r$ under the isomorphism

$$\sigma : J_\infty \to (\mathbf{R}^*)^{r_1} \times (\mathbf{C}^*)^{r_2} = (\{\pm 1\} \times \mathbf{C}_1^{r_2}) \times (\mathbf{R}^+)^r.$$

Then

$$J_\infty = U_\infty \times J'_\infty, \quad J' = U_0 \times J_\infty = U \times J'_\infty.$$

We denote by μ_{U_∞} and $\mu_{J'_\infty}$ the translate of the standard measures on $(\{\pm 1\}^{r_1} \times \mathbf{C}_1^{r_2})$ and $(\mathbf{R}^+)^r$ respectively under the above isomorphism. Then $\mu_{U_\infty}(U_\infty) = 2^{r_1}(4\pi)^{r_2}$, and

$$\mu_{J_\infty} = \mu_{U_\infty} \times \mu_{J'_\infty},$$

because μ_{J_∞} is, by definition, the translate of the standard measure on $(\mathbf{R}^*)^{r_1} \times (\mathbf{C}^*)^{r_2}$ under the same isomorphism. Put

$$\mu_U = \mu_{U_0} \times \mu_{U_\infty}.$$

Then

$$\mu_{J'} = \mu_{U_0} \times \mu_{J_\infty} = \mu_{U_0} \times \mu_{U_\infty} \times \mu_{J'_\infty} = \mu_U \times \mu_{J'_\infty}.$$

Let μ_W be the point measure of W so that $w = \mu_W(W) = [W : 1]$. With these $\mu_{J'}$, $\mu_{E\times T}$, μ_U, and μ_W, we apply the lemma to the diagram (B) and fix the measures of the groups which appear in that diagram. Then $\mu_{J'/(E\times T)} = \mu_{J'}/\mu_{E\times T}$ coincides with the measure of $J'/(E \times T)$ already fixed in (A). Since $\mu_{U/W}(U/W) = \mu_U(U)/\mu_W(W) = \mu_{U_0}(U_0)\mu_{U_\infty}(U_\infty)/\mu_W(W) = 2^{r_1}(2\pi)^{r_2}/w$, we have

$$\begin{aligned}
\mu(J'/(E \times T)) &= \mu(J'/(E \times T)U)\mu((E \times T)U/(E \times T)) \\
&= \mu(J'/(E \times T)U)\mu(U/W) \\
&= 2^{r_1}(4\pi)^{r_2} w^{-1} \mu(J'/(E \times T)U).
\end{aligned}$$

Now, the isomorphism $J'/U \to \mathbf{R}^r$ considered in the proof of Theorem 3.2 is the composite of the isomorphisms $J'/U \to J'_\infty \to (\mathbf{R}^+)^r \xrightarrow{\log} \mathbf{R}^r$. Since $\mu_{J'} = \mu_U \times \mu_{J'_\infty}$, we have $\mu_{J'/U} = \mu_{J'}/\mu_U \approx \mu_{J'_\infty}$. Also, the log map translates the standard measure of $(\mathbf{R}^+)^r$ to the standard measure $\mu_{\mathbf{R}^r}$ of $\mathbf{R}^r$. It follows that the translate of $\mu_{J'/U}$ under the isomorphism $J'/U \to \mathbf{R}^r$ is the standard measure $\mu_{\mathbf{R}^r}$ of $\mathbf{R}^r$. Let

$$\mu_{J'/(E\times T)U} \approx \mu_{\mathbf{R}^r/(\lambda(E)\oplus L)},$$
$$\mu_{(E\times T)U/U} \approx \mu_{\lambda(E)\oplus L},$$

under the isomorphisms induced by $J'/U \to \mathbf{R}^r$. Since $\mu_{J'/U} = \mu_{J'/(E\times T)U} \cdot \mu_{(E\times T)U/U}$, we have

$$\mu_{\mathbf{R}^r} = \mu_{\mathbf{R}^r/(\lambda(E)\oplus L)} \cdot \mu_{\lambda(E)\oplus L}.$$

Since $\mu_{(E\times T)U/U} \approx \mu_{(E\times T)/W} = (\mu_E \times \mu_T)/(\mu_W \times \mu_1) \approx \mu_E/\mu_W \times \mu_T$, we see that

$$\mu_{\lambda(E)\oplus L} = \mu_{\lambda(E)} \times \mu_L,$$

where $\mu_{\lambda(E)}$ is the point measure of $\lambda(E)$, and μ_L is the translate of μ_T under $T \to L$, namely, the measure defined by $d\mu_L(xw) = dx$, $x \in \mathbf{R}$.

Now, let $\varepsilon_1, \ldots, \varepsilon_{r-1}$ be units in E such that $\omega_1 = \lambda(\varepsilon_1), \ldots, \omega_{r-1} = \lambda(\varepsilon_{r-1})$ form a basis of the lattice $\lambda(E)$ over $\mathbf{Z}$. Then $\omega_1, \ldots, \omega_{r-1}, \omega_r$ form a basis of $\mathbf{R}^r$ over $\mathbf{R}$. Let

$$d\mu'_{\mathbf{R}^r}(\xi) = dx_1 \cdots dx_r$$

for $\xi = x_1\omega_1 + \cdots + x_r\omega_r$ in $\mathbf{R}^r$. Then $\mu'_{\mathbf{R}^r}$ is a Haar measure of $\mathbf{R}^r$, and $\mu_{\mathbf{R}^r} = c\mu'_{\mathbf{R}}$ for some $c > 0$. We see immediately that

$$(\mu'_{\mathbf{R}^r}/\mu_{\lambda(E)\oplus L})(\mathbf{R}^r/(\lambda(E) \oplus L)) = 1.$$

Hence

$$\begin{aligned}\mu_{\mathbf{R}^r/(\lambda(E)\oplus L)}(\mathbf{R}^r/(\lambda(E) \oplus L)) &= (\mu_{\mathbf{R}^r}/\mu_{\lambda(E)\oplus L})(\mathbf{R}^r/(\lambda(E) \oplus L)) \\ &= c \cdot (\mu'_{\mathbf{R}^r}/\mu_{\lambda(E)\oplus L})(\mathbf{R}^r/(\lambda(E) \oplus L)) \\ &= c,\end{aligned}$$

and consequently,

$$\begin{aligned}&\mu(J'/(E \times T)U) = \mu_{\mathbf{R}^r/(\lambda(E)\oplus L)}(\mathbf{R}^r/(\lambda(E) \oplus L)) = c, \\ &\mu(J'/(E \times T)) = 2^{r_1}(4\pi)^{r_2}w^{-1}c, \\ &v = h2^{r_1}(4\pi)^{r_2}w^{-1}c.\end{aligned}$$

Now, let $\xi = x_1\omega_1 + \cdots + x_r\omega_r = (y_1, \ldots, y_r)$. Since $\omega_k = (\log|\varepsilon_k^{(1)}|, \ldots, \log|\varepsilon_k^{(r)}|)$, $1 \le k \le r-1$, we have

$$y_j = \sum_{k=1}^{r-1} x_k \log|\varepsilon_k^{(j)}| + x_r \cdot \frac{1}{n}, \qquad j = 1, \ldots, r.$$

As $d\mu_{\mathbf{R}^r}(\xi) = dy_1 \cdots dy_r$ and $d\mu'_{\mathbf{R}}(\xi) = dx_1 \cdots dx_r$, we see that

$$\begin{aligned} c &= \left|\det\begin{pmatrix} \log|\varepsilon_1^{(1)}| & \cdots & \log|\varepsilon_1^{(r)}| \\ \vdots & & \vdots \\ \log|\varepsilon_{r-1}^{(1)}| & \cdots & \log|\varepsilon_{r-1}^{(r)}| \\ \frac{1}{n} & \cdots & \frac{1}{n}\end{pmatrix}\right| \\ &= 2^{-r_2}R,\end{aligned}$$

where

$$R = \left|\det\begin{pmatrix} e_1\log|\varepsilon_1^{(1)}| & \cdots & e_r\log|\varepsilon_1^{(r)}| \\ \vdots & & \vdots \\ e_1\log|\varepsilon_{r-1}^{(1)}| & \cdots & e_r\log|\varepsilon_{r-1}^{(r)}| \\ \frac{e_1}{n} & \cdots & \frac{e_r}{n}\end{pmatrix}\right|.$$

Since $R = 2^{r_2}c = 2^{r_2}\mu(J'/(E \times T)U)$, R is independent of the choice of $\varepsilon_1, \ldots, \varepsilon_{r-1}$ (such that $\omega_1, \ldots, \omega_{r-1}$ form a basis of $\lambda(E)$), and it is an invariant of the field F. We call it the regulator of F. If $r > 1$, then

$$R = \left|\det\begin{pmatrix} e_1 \log|\varepsilon_1^{(1)}| & \cdots & e_{r-1}\log|\varepsilon_1^{(r-1)}| & 0 \\ \vdots & & \vdots & 0 \\ e_1 \log|\varepsilon_{r-1}^{(1)}| & \cdots & e_{r-1}\log|\varepsilon_{r-1}^{(r-1)}| & 0 \\ \frac{e_1}{n} & \cdots & \frac{e_{r-1}}{n} & 1 \end{pmatrix}\right|$$

$$= \left|\det\begin{pmatrix} e_1 \log|\varepsilon_1^{(1)}| & \cdots & e_{r-1}\log|\varepsilon_1^{(r-1)}| \\ \vdots & & \vdots \\ e_1 \log|\varepsilon_{r-1}^{(1)}| & \cdots & e_{r-1}\log|\varepsilon_{r-1}^{(r-1)}| \end{pmatrix}\right|,$$

because $|N(\varepsilon)| = 1$ for any unit ε in E.

From the above computation and from Theorem 3.1, we now have the following theorem.

Theorem 3.3 *The measure $\mu_{\overline{J}_1}(\overline{J}_1)$ is given by*

$$v = h2^{r_1+r_2}\pi^{r_2}R/w,$$

and the residue of $\zeta_F(s)$ at $s = 1$ is equal to

$$\frac{2^{r_1+r_2}\pi^{r_2}R}{w\sqrt{d}}h.$$

Here h is the class number of F, R is the regulator of F, w is the number of roots of unity in F, and d is the absolute value of the discriminant of F.

Example 3.1 Let $F = \mathbf{Q}$. Then $h = 1$. Since $r = r_1 + r_2 = 1 + 0 = 1$, $R = |\frac{e_1}{n}| = 1$. We also see immediately that $w = 2$ ($W = \{\pm 1\}$) and $d = 1$. Hence the residue of $\zeta(s) = \zeta_{\mathbf{Q}}(s)$ at $s = 1$ is 1. However, in this special case, we can compute the residue also as follows. Let $s > 1$. Then

$$\int_1^\infty \frac{dx}{x^s} < \sum_{m=1}^\infty \frac{1}{m^s} < 1 + \int_1^\infty \frac{dx}{x^s},$$

namely,

$$\frac{1}{s-1} < \zeta(s) < 1 + \frac{1}{s-1}.$$

Hence

$$1 < (s-1)\zeta(s) < s,$$

and

$$\lim_{s\to 1}(s-1)\zeta(s) = 1.$$

Therefore the residue of $\zeta(s)$ at $s = 1$ is 1.

Now, let a denote the residue of $\zeta_F(s)$ at $s = 1$. Then Theorem 3.3 shows that

$$h = \frac{w\sqrt{d}a}{2^{r_1+r_2}\pi^{r_2}R}.$$

Hence, if we can complete the value of a by a different method, we obtain a formula which gives us the value of the class number h of F. Most of the classical class number formulae for algebraic number fields are obtained by this principle.

Chapter 4
Some Applications

4.1 Hecke Characters and Ideal Characters

Let $\mathfrak{m}$ be an integral ideal of F, $\mathfrak{m} = \prod_{\mathfrak{p}} \mathfrak{p}^{t_\mathfrak{p}} \neq \{0\}$, $t_\mathfrak{p} \geq 0$. Let $\mathfrak{p}_1, \ldots, \mathfrak{p}_g$ be all the prime ideals such that $t_\mathfrak{p} = \nu_\mathfrak{p}(\mathfrak{m}) > 0$. In Chap. 2, we defined the subgroups $J_0(\mathfrak{m})$ and $U_0(\mathfrak{m})$ of J by

$$J_0(\mathfrak{m}) = \{\alpha \mid \alpha = (a_P) \in J_0, \nu_{\mathfrak{p}_j}(a_{\mathfrak{p}_j} - 1) \geq t_{\mathfrak{p}_j}, 1 \leq j \leq g\},$$
$$U_0(\mathfrak{m}) = J_0(\mathfrak{m}) \cap U_0 = J_0(\mathfrak{m}) \cap U \quad (a_\mathfrak{p} = a_{P_\mathfrak{p}}).$$

Let $\mathfrak{I}(\mathfrak{m})$ be as before the subgroup of $\mathfrak{I}$ consisting of all ideals in $\mathfrak{I}$ which are prime to $\mathfrak{m}$. Then the map $\iota : J \to \mathfrak{I}$ induces a surjective homomorphism $J_0(\mathfrak{m}) \to \mathfrak{I}(\mathfrak{m})$ with kernel $U_0(\mathfrak{m})$ so that we obtain an isomorphism

$$J_0(\mathfrak{m})/U_0(\mathfrak{m}) \overset{\sim}{\longrightarrow} \mathfrak{I}(\mathfrak{m}).$$

Now, let χ be a Hecke character of F with conductor $\mathfrak{f}$; $\mathfrak{f}$ is the largest integral ideal of F such that $\chi(U_0(\mathfrak{f})) = 1$. It follows from the above that χ defines a character $\psi = \psi_\chi$ of $\mathfrak{I}(\mathfrak{f})$. Namely, given any ideal $\mathfrak{a}$ of $\mathfrak{I}(\mathfrak{f})$, we define

$$\psi(\mathfrak{a}) = \chi(\alpha),$$

where α is an idèle of $J_0(\mathfrak{f})$ such that $\mathfrak{a} = \iota(\alpha)$. For an integral ideal $\mathfrak{a}$ of F, we extend the definition of ψ by putting

$$\psi(\mathfrak{a}) = 0,$$

whenever $(\mathfrak{a}, \mathfrak{f}) \neq 1$. Thus $\psi(\mathfrak{a})$ is defined for $\mathfrak{a}$ which is either prime to $\mathfrak{f}$ or integral. In Chap. 3, we denoted the above (generalized) character of ideals by $\chi'(\mathfrak{a})$ or $\chi(\mathfrak{a})$. However, in the following, we shall denote it by ψ or ψ_χ to distinguish clearly from the Hecke character χ. By the definition,

© The Author(s), under exclusive license to Springer Nature Singapore Pte Ltd. 2019   67
K. Iwasawa, *Hecke's L-functions*, SpringerBriefs in Mathematics,
https://doi.org/10.1007/978-981-13-9495-9_4

$$L(s;\chi) = \prod_{\mathfrak{p}} (1 - \psi_\chi(\mathfrak{p}) N(\mathfrak{p})^{-s})^{-1}$$
$$= \sum_{\mathfrak{a} \subset \mathfrak{o}} \psi_\chi(\mathfrak{a}) N(\mathfrak{a})^{-s}, \qquad (\mathrm{Re}\,(s) > 1)$$

Let X denote the group of all Hecke characters of F. (Since X is the group of all characters of $J/F^* \times T$, it is a locally compact abelian group in Pontryagin topology. However, in the following we shall not be interested in such a topology on X.) Let $\mathfrak{m}$ be any non-zero integral ideal of F, and let $X(\mathfrak{m})$ denote the set of all Hecke characters χ in X such that $\chi(U_0(\mathfrak{m})) = 1$, namely, such that $\mathfrak{m}$ is contained in the conductor $\mathfrak{f}$ of χ; $\mathfrak{m} \subset \mathfrak{f}$, $\mathfrak{f}|\mathfrak{m}$. $X(\mathfrak{m})$ is obviously a subgroup of X. If $\mathfrak{m}_1 \subset \mathfrak{m}_2$ then $U_0(\mathfrak{m}_1) \subset U_0(\mathfrak{m}_2)$, and $X(\mathfrak{m}_2) \subset X(\mathfrak{m}_1)$. Clearly, $X = \bigcup_{\mathfrak{m}} X(\mathfrak{m})$.

Let $\chi \in X(\mathfrak{m})$ so that $\mathfrak{m} \subset \mathfrak{f} =$ conductor of χ. Let ψ_χ be the character of $\mathfrak{I}(\mathfrak{f})$ corresponding to χ. Since $\mathfrak{I}(\mathfrak{m}) \subset \mathfrak{I}(\mathfrak{f})$, ψ_χ induces a character $\psi_\chi^{\mathfrak{m}}$ of $\mathfrak{I}(\mathfrak{m})$. Let $\mathrm{Ch}(\mathfrak{I}(\mathfrak{m}))$ denote the group of all characters of $\mathfrak{I}(\mathfrak{m})$. Then $\chi \mapsto \psi_\chi^{\mathfrak{m}}$ obviously defines a homomorphism

$$X(\mathfrak{m}) \longrightarrow \mathrm{Ch}(\mathfrak{I}(\mathfrak{m})).$$

Proposition 4.1 *The above homomorphism is injective.*

Proof We first prove the following lemma.

Lemma 4.1 *The group $J_0(\mathfrak{m})F^*$ is everywhere dense in J.*

Proof Let $P_1, \ldots, P_h$ consist of all infinite prime spots of F and of all finite spots of F such that $\nu_P(\mathfrak{m}) > 0$. Let $\alpha = (a_P)$ be any idèle in J, and let $0 < \varepsilon < 1$. By the approximation theorem, there exists an element a in F such that

$$\nu_P(a_P - a) < \varepsilon \nu_P(a_P),$$

for $P = P_1, \ldots, P_h$. Since $\varepsilon < 1$, we know $a \neq 0$, so $a \in F^*$. Let $\alpha' = (a'_P)$ with $a'_P = 1$ for $P = P_1, \ldots, P_h$ and $a'_P = a^{-1}a_P$ for $P \neq P_1, \ldots, P_h$. Let $\alpha'' = \alpha' a = (a''_P)$. Then $\alpha' \in J_0(\mathfrak{m})$, $\alpha'' \in J_0(\mathfrak{m})F^*$, $a''_P = a_P$ for $P = P_1, \ldots, P_h$, and $\nu_P(1 - a''_P a_P^{-1}) < \varepsilon$ for $P = P_1, \ldots, P_h$. With small $\varepsilon > 0$, α'' is then very close to α. Hence $J_0(\mathfrak{m})F^*$ is dense in J.

Now, suppose that $\psi_\chi^{\mathfrak{m}} = 1$ for a character χ in $X(\mathfrak{m})$. Then $\chi(J_0(\mathfrak{m})) = \psi_\chi^{\mathfrak{m}}(\mathfrak{I}(\mathfrak{m})) = 1$. By the definition of a Hecke character, $\chi(F^*) = 1$. Hence $\chi(J_0(\mathfrak{m})F^*) = 1$. Since χ is continuous, it follows from the above lemma that $\chi(J) = 1$, namely, $\chi \equiv 1$.

Remark 4.1 $X(\mathfrak{m}) \to \mathrm{Ch}(\mathfrak{I}(\mathfrak{m}))$ is not surjective.

Let $\sigma : J_\infty \to \mathbf{R}^* \times \cdots \times \mathbf{R}^* \times \mathbf{C}^* \times \cdots \times \mathbf{C}^*$ be defined as before, and let

$$J_\infty^0 = \sigma^{-1}(\mathbf{R}^+ \times \cdots \times \mathbf{R}^+ \times \mathbf{C}^* \times \cdots \mathbf{C}^*).$$

J_∞^0 is the set of all $\alpha = (a_P)$ in J_∞ such that $\sigma_k(a_{P_{\infty,k}}) > 0$ for $k = 1, \ldots, r_1$. It is an open subgroup of index 2^{r_1} in J_∞, and it contains the subgroup T; $T \subset J_\infty^0$.

Problem 4.1 Prove that J_∞^0 is the connected component of 1 in J (and in J_∞).

Let X' denote the subgroup of all characters χ in X such that $\chi(J_\infty^0) = 1$. Let $\{m_1, \ldots, m_n; s_1, \ldots, s_n\}$ be the signature of a character χ of X. Then χ belongs to X' if and only if $m_k = 0$ for $r_1 + 1 \leq k \leq n$, and $s_k = 0$ for $1 \leq k \leq n$.

For any non-zero integral ideal $\mathfrak{m}$, we put $X'(\mathfrak{m}) = X' \cap X(\mathfrak{m})$. Then $X'(\mathfrak{m}_2) \subset X'(\mathfrak{m}_1)$ for $\mathfrak{m}_1 \subset \mathfrak{m}_2$, and $X' = \bigcup_{\mathfrak{m}} X'(\mathfrak{m})$. We shall next determine the image of $X'(\mathfrak{m})$ under $X(\mathfrak{m}) \to \mathrm{Ch}(\mathfrak{I}(\mathfrak{m}))$.

Let

$$F(\mathfrak{m}) = F^* \cap (J_0(\mathfrak{m}) \times J_\infty^0);$$

$F(\mathfrak{m})$ is the subgroup of all a in F; $a \neq 0$, such that

$$\begin{aligned} &\nu_\mathfrak{p}(a - 1) \geq t_\mathfrak{p} = \nu_\mathfrak{p}(\mathfrak{m}), && \text{for } \nu_\mathfrak{p}(\mathfrak{m}) > 0, \\ &\sigma_k(a) > 0 && \text{for } 1 \leq k \leq r_1. \end{aligned}$$

Lemma 4.2 *There exists a natural isomorphism*

$$J/F^*(U_0(\mathfrak{m}) \times J_\infty^0) \longrightarrow J_0(\mathfrak{m}) \times J_\infty^0/F(\mathfrak{m})(U_0(\mathfrak{m}) \times J_\infty^0),$$

and the both groups are finite.

Proof Since $J_0(\mathfrak{m})$ ($U_0(\mathfrak{m})$) is open in J_0, and since J_∞^0 is open in J_∞, $J_0(\mathfrak{m}) \times J_\infty^0$ ($U_0(\mathfrak{m}) \times J_\infty^0$) is open in $J = J_0 \times J_\infty$. By Lemma 4.1, $F^*(J_0(\mathfrak{m}) \times J_\infty^0)$ is everywhere dense in J. Hence $J = F^*(J_0(\mathfrak{m}) \times J_\infty^0) = F^*(U_0(\mathfrak{m}) \times J_\infty^0)(J_0(\mathfrak{m}) \times J_\infty^0)$. On the other hand, $F^*(U_0(\mathfrak{m}) \times J_\infty^0) \cap (J_0(\mathfrak{m}) \times J_\infty^0) = (F^* \cap (J_0(\mathfrak{m}) \times J_\infty^0)) \cdot (U_0(\mathfrak{m}) \times J_\infty^0) = F(\mathfrak{m})(U_0(\mathfrak{m}) \times J_\infty^0)$. Hence we have the above isomorphism. Since $U_0(\mathfrak{m}) \times J_\infty^0$ is open in J, $J/F^*(U_0(\mathfrak{m}) \times J_\infty^0)$ is discrete. However, $F^* \times T \subset F^*(U_0(\mathfrak{m}) \times J_\infty^0)$, and $J/F^* \times T$ is compact (Theorem 3.2). Hence $J/F^*(U_0(\mathfrak{m}) \times J_\infty^0)$ is also compact. It then follows that $J/F^*(U_0(\mathfrak{m}) \times J_\infty^0)$ is a finite group.

Now, let

$$\mathfrak{r}(\mathfrak{m}) = \iota(F(\mathfrak{m}));$$

$\mathfrak{r}(\mathfrak{m})$ is the group of all principal ideals (a) with a in $F(\mathfrak{m})$. Since $\iota(J_0(\mathfrak{m}) \times J_\infty^0) = \iota(J_0(\mathfrak{m})) = \mathfrak{I}(\mathfrak{m})$, and since the kernel of $J_0(\mathfrak{m}) \times J_\infty^0 \to \mathfrak{I}(\mathfrak{m})$ is $U_0(\mathfrak{m}) \times J_\infty^0$, we see that $\mathfrak{r}(\mathfrak{m})$ is a subgroup of $\mathfrak{I}(\mathfrak{m})$, and that the isomorphism $J_0(\mathfrak{m}) \times J_\infty^0/U_0(\mathfrak{m}) \times J_\infty^0 \cong \mathfrak{I}(\mathfrak{m})$ induces isomorphisms

$$\begin{aligned} F(\mathfrak{m})(U_0(\mathfrak{m}) \times J_\infty^0)/U_0(\mathfrak{m}) \times J_\infty^0 &\longrightarrow \mathfrak{r}(\mathfrak{m}), \\ J_0(\mathfrak{m}) \times J_\infty^0/F(\mathfrak{m})(U_0(\mathfrak{m}) \times J_\infty^0) &\longrightarrow \mathfrak{I}(\mathfrak{m})/\mathfrak{r}(\mathfrak{m}). \end{aligned}$$

It follows in particular that $\mathfrak{I}(\mathfrak{m})/\mathfrak{r}(\mathfrak{m})$ is a finite group.

Proposition 4.2 *Under the homomorphism in Proposition 4.1, $X'(\mathfrak{m})$ is mapped to the group of all characters of $\mathfrak{J}(\mathfrak{m})$ which are trivial on $\mathfrak{r}(\mathfrak{m})$. Hence*

$$X'(\mathfrak{m}) \cong \mathrm{Ch}(\mathfrak{J}(\mathfrak{m})/\mathfrak{r}(\mathfrak{m})).$$

Proof Let χ be a Hecke character of F. Then $\chi(F^* \times T) = 1$. By the definition, χ belongs to $X'(\mathfrak{m})$ if and only if $\chi(U_0(\mathfrak{m}) \times J_\infty^0) = 1$. Since $T \subset J_\infty^0$, we see that $X'(\mathfrak{m})$ is the group of all characters of J such that $\chi(F^*(U_0(\mathfrak{m}) \times J_\infty^0)) = 1$. It then follows from the above isomorphisms

$$J/F^*(U_0(\mathfrak{m}) \times J_\infty^0) \longrightarrow J_0(\mathfrak{m}) \times J_\infty^0/F(\mathfrak{m})(U_0(\mathfrak{m}) \times J_\infty^0) \longrightarrow \mathfrak{J}(\mathfrak{m})/\mathfrak{r}(\mathfrak{m})$$

that when χ ranges over $X'(\mathfrak{m})$, $\psi_\chi^{\mathfrak{m}}$ gives us all characters of $\mathfrak{J}(\mathfrak{m})$ that are trivial on $\mathfrak{r}(\mathfrak{m})$.

Proposition 4.3 *X' is the subgroup of all elements of finite order in X.*

Proof Since $\mathfrak{J}(\mathfrak{m})/\mathfrak{r}(\mathfrak{m})$ is a finite group, so are $\mathrm{Ch}(\mathfrak{J}(\mathfrak{m})/\mathfrak{r}(\mathfrak{m}))$ and $X'(\mathfrak{m})$. Hence every element in $X' = \bigcup X'(\mathfrak{m})$ has a finite order. Suppose conversely that χ has a finite order in X. The restriction χ' of χ on J_∞^0 then also has a finite order. However, as $J_\infty^0 \cong \mathbf{R}^+ \times \cdots \times \mathbf{R}^+ \times \mathbf{C}^* \times \cdots \times \mathbf{C}^*$, no non-trivial character of J_∞^0 has a finite order. Hence $\chi' = 1$, $\chi(J_\infty^0) = 1$, and χ' belongs to X'.

Let $\mathfrak{m}_1 \subset \mathfrak{m}_2$, both non-zero integral ideals of F. Then $X'(\mathfrak{m}_2) \subset X'(\mathfrak{m}_1)$, $\mathfrak{J}(\mathfrak{m}_1) \subset \mathfrak{J}(\mathfrak{m}_2)$, and $\mathfrak{r}(\mathfrak{m}_1) \subset \mathfrak{r}(\mathfrak{m}_2)$. Hence we have homomorphisms $X'(\mathfrak{m}_2) \to X'(\mathfrak{m}_1)$, $\mathfrak{J}(\mathfrak{m}_1) \to \mathfrak{J}(\mathfrak{m}_2)$, $\mathfrak{J}(\mathfrak{m}_1)/\mathfrak{r}(\mathfrak{m}_1) \to \mathfrak{J}(\mathfrak{m}_2)/\mathfrak{r}(\mathfrak{m}_2)$, and $\mathrm{Ch}(\mathfrak{J}(\mathfrak{m}_2)/\mathfrak{r}(\mathfrak{m}_2)) \to \mathrm{Ch}(\mathfrak{J}(\mathfrak{m}_1)/\mathfrak{r}(\mathfrak{m}_1))$. Using Proposition 4.2, we then obtain the following commutative diagram:

$$\begin{array}{ccc}
X'(\mathfrak{m}_2) & \longrightarrow & X'(\mathfrak{m}_1) \\
\downarrow & & \downarrow \\
\mathrm{Ch}(\mathfrak{J}(\mathfrak{m}_2)/\mathfrak{J}(\mathfrak{m}_2)) & \longrightarrow & \mathrm{Ch}(\mathfrak{J}(\mathfrak{m}_1)/\mathfrak{J}(\mathfrak{m}_1)).
\end{array}$$

Since the map in the upper level is injective, so is the map in the lower level.

Let χ be a character in $X'(\mathfrak{m})$ with conductor $\mathfrak{f}$. Then $\mathfrak{m} \subset \mathfrak{f}$, and $\mathfrak{f}$ is the largest integral ideal $\neq \{0\}$ with the property $\chi \in X'(\mathfrak{f})$. Hence $\mathfrak{f} = \mathfrak{m}$ if and only if $\chi \notin X'(\mathfrak{m}')$ for any $\mathfrak{m}' \neq \mathfrak{m}$, $\mathfrak{m} \subset \mathfrak{m}' \subset \mathfrak{o}$.

A character in $\mathrm{Ch}(\mathfrak{J}(\mathfrak{m})/\mathfrak{r}(\mathfrak{m}))$ (namely, a character of $\mathfrak{J}(\mathfrak{m})$ which is trivial on $\mathfrak{r}(\mathfrak{m})$) is called primitive if it is not in the image of $\mathrm{Ch}(\mathfrak{J}(\mathfrak{m}')/\mathfrak{r}(\mathfrak{m}')) \to \mathrm{Ch}(\mathfrak{J}(\mathfrak{m})/\mathfrak{r}(\mathfrak{m}))$ for any $\mathfrak{m}' \neq \mathfrak{m}$, $\mathfrak{m} \subset \mathfrak{m}' \subset \mathfrak{o}$. It follows from the above that a character χ in $X'(\mathfrak{m})$ has conductor $\mathfrak{m}$ if and only if the corresponding $\psi_\chi^{\mathfrak{m}}$ is primitive in $\mathrm{Ch}(\mathfrak{J}(\mathfrak{m})/\mathfrak{r}(\mathfrak{m}))$. In such a case, $\psi_\chi(\mathfrak{a}) = \psi_\chi^{\mathfrak{m}}(\mathfrak{a})$ for $\mathfrak{a}$ prime to $\mathfrak{f} = \mathfrak{m}$, and $\psi_\chi(\mathfrak{a}) = 0$ for integral $\mathfrak{a}$ with $(\mathfrak{a}, \mathfrak{f}) \neq 1$. Therefore in order to obtain all L-functions $L(s; \chi) = \prod(1 - \psi_\chi(\mathfrak{p})N(\mathfrak{p})^{-s})^{-1}$ for the characters χ in X' with conductor $\mathfrak{m}$, it is sufficient to find all primitive characters of the finite group $\mathfrak{J}(\mathfrak{m})/\mathfrak{r}(\mathfrak{m})$.

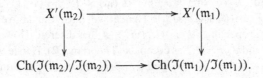

Let $\overline{F^* J_\infty^0}$ denote the closure of $F^* J_\infty^0$ in J. Since $F^* \times T \subset \overline{F^* J_\infty^0}$, $J/\overline{F^* J_\infty^0}$ is a compact abelian group. By the definition of X', X' may be considered as the group of all characters of $J/\overline{F^* J_\infty^0}$. Let A be the maximal abelian extension of F in $\mathbf{C}$, and let G be the Galois group of A/F; G is a compact abelian group in Krull topology. One of the fundamental theorems in class field theory states that $J/\overline{F^* J_\infty^0}$ is topologically isomorphic to G in a natural manner, or equivalently, that X' is naturally isomorphic to the group of all characters of the compact abelian group G. It was C. Chevalley who first introduced the groups J, X, X' etc., and simplified (clarified) the classical class field theory which had been based upon the groups $\mathfrak{I}(m)/\mathfrak{r}(\mathfrak{m})$ and $\mathrm{Ch}(\mathfrak{I}(\mathfrak{m})/\mathfrak{r}(\mathfrak{m}))$, $\mathfrak{m} \subset \mathfrak{o}$.

4.2 The Existence of Prime Ideals

Let χ be a Hecke character of F. Since

$$L(s;\chi) = \prod_{\mathfrak{p}} \left(1 - \psi_\chi(\mathfrak{p}) N(\mathfrak{p})^{-s}\right)^{-1}, \ \mathrm{Re}\,(s) > 1,$$

we obtain

$$\begin{aligned} \log L(s;\chi) &= -\sum_{\mathfrak{p}} \log\left(1 - \psi_\chi(\mathfrak{p}) N(\mathfrak{p})^{-s}\right) \\ &= \sum_{\mathfrak{p}} \sum_{m=1}^{\infty} \frac{1}{m} \psi_\chi(\mathfrak{p})^m N(\mathfrak{p})^{-ms}, \ \mathrm{Re}\,(s) > 1. \end{aligned} \tag{4.1}$$

Here $\log z$ denotes the branch of the log function such that $\log 1 = 0$. The double sum in the above is absolutely convergent, and we see that

$$L(s;\chi) = \exp\left(\sum_{\mathfrak{p}} \sum_{m=1}^{\infty} \frac{1}{m} \psi_\chi(\mathfrak{p})^m N(\mathfrak{p})^{-ms} \right), \ \mathrm{Re}\,(s) > 1.$$

Lemma 4.3 *For any real ε, y, $\varepsilon > 0$,*

$$\left| L(1+\varepsilon;1)^3 L(1+\varepsilon+iy;\chi)^4 L(1+\varepsilon+2iy;\chi^2) \right| \geq 1.$$

Proof Let A denote the product of the L-functions in the above. Then it follows from (4.1) that

$$A = \exp\left(\sum_{\mathfrak{p}} \sum_{m=1}^{\infty} a(\mathfrak{p}, m) \right),$$

with

$$a(\mathfrak{p}, m) = \frac{1}{m}\left(3 + 4\psi_\chi(\mathfrak{p})^m N(\mathfrak{p})^{-imy} + \psi_\chi(\mathfrak{p})^{2m} N(\mathfrak{p})^{-2imy}\right) N(\mathfrak{p})^{-m(1+\varepsilon)}.$$

If $\psi_\chi(\mathfrak{p}) = 0$, then $\mathrm{Re}\,(a(\mathfrak{p}, m)) = \frac{3}{m} N(\mathfrak{p})^{-m(1+\varepsilon)} > 0$. If $\psi_\chi(\mathfrak{p}) \neq 0$, then $\left|\psi_\chi(\mathfrak{p})^m N(\mathfrak{p})^{-imy}\right| = 1$. Hence, let $\psi_\chi(\mathfrak{p})^m N(\mathfrak{p})^{-imy} = e^{i\theta}$, $\theta \in \mathbf{R}$. Then

$$\begin{aligned}
\mathrm{Re}\,(a(\mathfrak{p}, m)) &= \frac{1}{m}(3 + 4\cos\theta + \cos 2\theta) N(\mathfrak{p})^{-m(1+\varepsilon)} \\
&= \frac{1}{m}(3 + 4\cos\theta + 2\cos^2\theta - 1) N(\mathfrak{p})^{-m(1+\varepsilon)} \\
&= \frac{2}{m}(1 + \cos\theta)^2 N(\mathfrak{p})^{-m(1+\varepsilon)} \\
&\geq 0.
\end{aligned}$$

It follows that

$$|A| = \exp\left(\sum_{\mathfrak{p}} \sum_{m} \mathrm{Re}\,(a(\mathfrak{p}, m))\right) \geq 1.$$

Proposition 4.4 *For any real y,*

$$L(1 + iy; \chi) \neq 0$$

Proof (1) Suppose that $\chi^2 \equiv 1$. Then $\psi_\chi(\mathfrak{p}) = 0$ or ± 1 for any prime ideal $\mathfrak{p}$, and

$$\begin{aligned}
&\left(1 - N(\mathfrak{p})^{-s}\right)^{-1} \left(1 - \psi_\chi(\mathfrak{p}) N(\mathfrak{p})^{-s}\right)^{-1} \\
&= \begin{cases}
\left(1 - N(\mathfrak{p})^{-s}\right)^{-1} = 1 + N(\mathfrak{p})^{-s} + N(\mathfrak{p})^{-2s} + \cdots, \\
\left(1 - N(\mathfrak{p})^{-s}\right)^{-2} = 1 + 2N(\mathfrak{p})^{-s} + 3N(\mathfrak{p})^{-2s} + \cdots, \\
\left(1 - N(\mathfrak{p})^{-2s}\right)^{-1} = 1 \qquad\qquad + N(\mathfrak{p})^{-2s} + \cdots.
\end{cases}
\end{aligned}$$

Let

$$\begin{aligned}
Z(s) &= L(s; 1) L(s; \chi), \quad (s \in \mathbf{C}), \\
&= \prod_{\mathfrak{p}} \left(1 - N(\mathfrak{p})^{-s}\right)^{-1} \left(1 - \psi_\chi(\mathfrak{p}) N(\mathfrak{p})^{-s}\right), \quad \mathrm{Re}\,(s) > 1, \\
&= \sum_{m=1}^{\infty} a_m m^{-s}, \quad \mathrm{Re}\,(s) > 1. \qquad (4.2)
\end{aligned}$$

Then we see from the above that $a_m \in \mathbf{Z}$, $a_m \geq 0$, and that $a_m \geq 1$ if $m = N(\mathfrak{p})^2$ for some prime ideal $\mathfrak{p}$. Hence, for any prime number p, there exists m such that m is a power of p and $a_m \geq 1$. It follows that

$$\sum_{m=1}^{\infty} a_m = +\infty.$$

Now, suppose that $L(1;\chi) = 0$. Since $L(s;1)$ has a unique simple pole at $s = 1$, (Theorem 3.1), $Z(s)$ is then holomorphic at any complex s. Hence

$$Z(s) = \sum_{l=0}^{\infty} \frac{Z^{(l)}(2)}{l!}(s-2)^l, \quad s \in \mathbf{C}$$

Here $Z^{(l)}(2)$ can be computed from (4.2):

$$Z^{(l)}(s) = \sum_{m=1}^{\infty} a_m(-\log m)^l m^{-s}, \quad \mathrm{Re}\,(s) > 1,$$

$$Z^{(l)}(2) = \sum_{m=1}^{\infty} a_m(-\log m)^l m^{-2}.$$

Using $a_m \geq 0$, we then have

$$\begin{aligned}
Z(2) &= \sum_{l=0}^{\infty} \frac{1}{l!}(-2)^l \sum_{m=1}^{\infty} a_m\,(-\log m)^l\, m^{-2} \\
&= \sum_{l=0}^{\infty} \left(\sum_{m=1}^{\infty} \frac{2^l}{l!} a_m\,(\log m)^l\, m^{-2} \right) \\
&= \sum_{m=1}^{\infty} \left(\sum_{l=0}^{\infty} \frac{1}{l!}\,(2\log m)^l \right) a_m m^{-2} \\
&= \sum_{m=1}^{\infty} e^{2\log m} a_m m^{-2} \\
&= \sum_{m=1}^{\infty} a_m \\
&= +\infty.
\end{aligned}$$

This contradicts the fact that $Z(s)$ is holomorphic at $s = 0$. Hence $L(1;\chi) \neq 0$.

(2) To prove the proposition, we may now assume that $y \neq 0$ if $\chi^2 \equiv 1$. Under such an assumption, $L(s;\chi^2)$ is always holomorphic at $s = 1 + 2iy$. Hence

$$\lim_{\varepsilon \to 0} L(1 + \varepsilon + 2iy;\chi^2) = L(1 + 2iy;\chi^2), \quad \varepsilon > 0.$$

Since $L(s; 1)$ has a simple pole at $s = 1$, we also have

$$\lim_{\varepsilon \to 0} \varepsilon L(1+\varepsilon; 1) = a = \text{residue of } L(s; 1) \text{ at } s = 1.$$

Suppose now that $L(1+iy; \chi) = 0$. Then $L(s; \chi)$ is holomorphic at $s = 1 + iy$, and

$$\begin{aligned}\lim_{\varepsilon \to 0} \frac{L(1+\varepsilon+iy; \chi)}{\varepsilon} &= \lim_{\varepsilon \to 0} \frac{L(1+\varepsilon+iy; \chi) - L(1+iy; \chi)}{\varepsilon} \\ &= L'(1+iy; \chi).\end{aligned}$$

It follows that

$$\begin{aligned}&\lim_{\varepsilon \to 0} L(1+\varepsilon; 1)^3 L(1+\varepsilon+iy; \chi)^4 L(1+\varepsilon+2iy; \chi^2) \\ &= \lim_{\varepsilon \to 0} (\varepsilon L(1+\varepsilon; 1))^3 \left(\frac{L(1+\varepsilon+iy; \chi)}{\varepsilon}\right)^4 L(1+\varepsilon+2iy; \chi^2)\varepsilon \\ &= 0.\end{aligned}$$

However, this contradicts the previous lemma. Therefore $L(1+iy; \chi) \neq 0$.

It follows from the proposition that $\xi(s; \chi) \neq 0$ on the line $\mathrm{Re}\,(s) = 2$. By the functional equation, we then have $\xi(s; \chi) \neq 0$ also on the line $\mathrm{Re}\,(s) = 0$. Therefore the zeros of $L(s; \chi)$ at $s \neq 0$ on $\mathrm{Re}\,(s) = 0$ are given by the zeros of

$$\gamma(s; \chi)^{-1} = \prod_{k=1}^{r} \Gamma\left(\frac{e_k}{2}\left(s + \frac{m_k + m_k'}{2} + is_k\right)\right)^{-1}$$

on the same line; at $s = 0$, we have to take into account the fact that $\xi(s; \chi)$ has a simple pole at $s = 0$. If $\mathrm{Re}\,(s) = 0$, then $\mathrm{Re}\,(\frac{e_k}{2}(s + \frac{m_k+m_k'}{2} + is_k)) = \frac{e_k}{4}(m_k + m_k')$. Hence $\Gamma(\frac{e_k}{2}(s + \frac{m_k+m_k'}{2} + is_k))^{-1}$ has zero on $\mathrm{Re}\,(s) = 0$ only when $m_k = m_k' = 0$, and in such a case, it has a unique simple zero at $s = -is_k$. Thus the zeros of $L(s; \chi)$ on $\mathrm{Re}\,(s) = 0$ are explicitly determined. For example, if $\chi \equiv 1$, then $m_k = m_k' = s_k = 0$, $1 \leq k \leq r$, and $\gamma(s; 1)^{-1}$ has a zero of order r at $s = 0$. Hence $\zeta_F(s)$ has a zero of order $r - 1$ $(r = r_1 + r_2)$ at $s = 0$, but $\zeta_F(s) \neq 0$ for $s \neq 0$ with $\mathrm{Re}\,(s) = 0$.

Let $\mathfrak{m}$ be any non-zero integral ideal of F. Let χ be a character in $X'(\mathfrak{m})$ with conductor $\mathfrak{f}$, $\psi = \psi_\chi$ the corresponding character of $\mathfrak{I}(\mathfrak{f})$ with $\psi(\mathfrak{a}) = 0$ for integral $\mathfrak{a}$, not prime to $\mathfrak{f}$, and $\psi_\chi^{\mathfrak{m}}$ the restriction of ψ_χ on $\mathfrak{I}(\mathfrak{m})$.

Let $\mathrm{Re}\,(s) \geq a > \frac{1}{2}$. Then

$$\sum_{\mathfrak{p}}\sum_{m=2}^{\infty}\left|\frac{1}{m}\psi(\mathfrak{p})^m N(\mathfrak{p})^{-ms}\right| \leq \sum_{\mathfrak{p}}\sum_{m=2}^{\infty} N(\mathfrak{p})^{-am}$$
$$= \sum_{\mathfrak{p}} \frac{N(\mathfrak{p})^{-2a}}{1-N(\mathfrak{p})^{-a}}$$
$$\leq \frac{1}{1-2^{-a}}\sum_{\mathfrak{p}} N(\mathfrak{p})^{-2a} < +\infty,$$

because $\sum_{\mathfrak{a}\subset\mathfrak{o}} N(\mathfrak{a})^{-2a} < +\infty$ for $2a > 1$. Hence $\sum_{\mathfrak{p}}\sum_{m=2}^{\infty}\frac{1}{m}\psi(\mathfrak{p})^m N(\mathfrak{p})^{-ms}$ is a holomorphic function of s for $\mathrm{Re}\,(s) > \frac{1}{2}$, and we obtain from (4.1) that

$$\log L(s;\chi) \sim \sum_{\mathfrak{p}} \psi_\chi^{\mathfrak{m}}(\mathfrak{p})N(\mathfrak{p})^{-s}.$$

Here and in the following, we shall use the notation $\sim$ to indicate the fact that the difference of the both sides of $\sim$ is holomorphic at $s = 1$. If $\mathfrak{p}$ is in $\mathfrak{J}(\mathfrak{m})$, namely, if $\mathfrak{p}\nmid\mathfrak{m}$, then $\psi_\chi(\mathfrak{p}) = \psi_\chi^{\mathfrak{m}}(\mathfrak{p})$. However, there exist only a finite number of $\mathfrak{p}$'s such that $\mathfrak{p}|\mathfrak{m}$. Hence it follows from the above that

$$\log L(s;\chi) \sim \sum_{\mathfrak{p}\in\mathfrak{J}(\mathfrak{m})} \psi_\chi^{\mathfrak{m}}(\mathfrak{p})N(\mathfrak{p})^{-s}.$$

Since χ is a character of $X'(\mathfrak{m})$, we know from Sect. 4.1 that $\psi_\chi^{\mathfrak{m}}(\mathfrak{r}(\mathfrak{m})) = 1$ so that $\psi_\chi^{\mathfrak{m}}$ may be considered as a character of $\mathfrak{J}(\mathfrak{m})/\mathfrak{r}(\mathfrak{m})$. Hence, for any c in $\mathfrak{J}(\mathfrak{m})/\mathfrak{r}(\mathfrak{m})$, $\psi_\chi^{\mathfrak{m}}(c)$ is defined. In particular, $\psi_\chi^{\mathfrak{m}}(c_{\mathfrak{p}}) = \psi_\chi^{\mathfrak{m}}(\mathfrak{p})$ if $c_{\mathfrak{p}}$ denote the coset of $\mathfrak{p}$ in $\mathfrak{J}(\mathfrak{m})/\mathfrak{r}(\mathfrak{m})$, and we have

$$\log L(s;\chi) \sim \sum_{\mathfrak{p}\in\mathfrak{J}(\mathfrak{m})} \psi_\chi^{\mathfrak{m}}(c_{\mathfrak{p}})N(\mathfrak{p})^{-s}. \tag{4.3}$$

Now, when χ ranges over all characters in $X'(\mathfrak{m})$, $\psi_\chi^{\mathfrak{m}}$ ranges over all characters of $\mathfrak{J}(\mathfrak{m})/\mathfrak{r}(\mathfrak{m})$ (Proposition 4.2). Therefore it follows from (4.3) that

$$\sum_{\chi\in X'(\mathfrak{m})} \overline{\psi}_\chi^{\mathfrak{m}}(c)\log L(s;\chi) \sim \sum_{\chi}\sum_{\mathfrak{p}\in\mathfrak{J}(\mathfrak{m})} \overline{\psi}_\chi^{\mathfrak{m}}(c)\psi_\chi^{\mathfrak{m}}(c_{\mathfrak{p}})N\mathfrak{p}^{-s}$$
$$= \sum_{\mathfrak{p}\in\mathfrak{J}(\mathfrak{m})}\sum_{\chi} \psi_\chi\psi_\chi^{\mathfrak{m}}(c^{-1}c_{\mathfrak{p}})N(\mathfrak{p})^{-s},$$
$$= \sum_{\mathfrak{p}\in\mathfrak{J}(\mathfrak{m})}\ \sum_{\psi\in\mathrm{Ch}(\mathfrak{J}(\mathfrak{m})/\mathfrak{r}(\mathfrak{m}))} \psi(c^{-1}c_{\mathfrak{p}})N(\mathfrak{p})^{-s},$$

for any c in $\mathfrak{J}(\mathfrak{m})/\mathfrak{r}(\mathfrak{m})$. Since $\mathfrak{J}(\mathfrak{m})/\mathfrak{r}(\mathfrak{m})$ is a finite group, we know that

$$\sum_{\psi \in \mathrm{Ch}(\mathfrak{I}(\mathfrak{m})/\mathfrak{r}(\mathfrak{m}))} \psi(c^{-1}c_{\mathfrak{p}}) = \begin{cases} N = [\mathfrak{I}(\mathfrak{m}) : \mathfrak{r}(\mathfrak{m})], & \text{if } c = c_{\mathfrak{p}}, \\ 0, & \text{if } c \neq c_{\mathfrak{p}}. \end{cases}$$

Hence

$$\sum_{\chi \in X'(\mathfrak{m})} \overline{\psi}_{\chi}^{\mathfrak{m}}(c) \log L(s; \chi) \sim N \sum_{\mathfrak{p} \in c} N(\mathfrak{p})^{-s}.$$

If $\chi \neq 1$, then $L(s; \chi)$ is holomorphic at $s = 1$ (Theorem 3.1), and $L(1; \chi) \neq 0$ by the above proposition. Hence

$$\log L(s; \chi) \sim 0.$$

On the other hand, if $\chi \equiv 1$, then $L(s; 1) = \zeta_F(s)$ has a simple pole at $s = 1$ (Theorem 3.1) : $L(s; 1) = (s-1)^{-1}(a_0 + a_1(s-1) + \cdots)$, $a_0 \neq 0$. It follows that

$$\log L(s; 1) \sim \log \frac{1}{s-1}.$$

Since $\psi_{\chi}^{\mathfrak{m}}(c) = 1$ for $\chi \equiv 1$, we obtain from the above the following theorem.

Theorem 4.1 *Let $\mathfrak{m}$ be any integral ideal of F. Let N be the order of $\mathfrak{I}(\mathfrak{m})/\mathfrak{r}(\mathfrak{m})$, and let c be any coset of $\mathfrak{I}(\mathfrak{m})/\mathfrak{r}(\mathfrak{m})$. Then*

$$\sum_{\mathfrak{p} \in c} N(\mathfrak{p})^{-s} \sim \frac{1}{N} \log \frac{1}{s-1}.$$

In particular, each coset of $\mathfrak{I}(\mathfrak{m})/\mathfrak{r}(\mathfrak{m})$ (i.e., each ideal class of the ideal class group $\mathfrak{I}(\mathfrak{m})/\mathfrak{r}(\mathfrak{m})$) contains infinitely many prime ideals in it.

Proof The first half follows immediately from the above. Let $s \to 1$, $s > 1$. Then $\log \frac{1}{s-1} \to +\infty$. Hence $\sum_{\mathfrak{p} \in c} N(\mathfrak{p})^{-s} \to +\infty$ also. Therefore there exist infinitely many $\mathfrak{p}$'s in c.

Note that for the proof of Theorem 4.1, it is sufficient to know that $L(1; \chi) \neq 0$.

Let $\mathfrak{m} = \mathfrak{o}$. Then $\mathfrak{I}(\mathfrak{o}) = \mathfrak{I}$, $\mathfrak{r}(\mathfrak{o}) \subset \mathfrak{H}$ so that $\mathfrak{I}/\mathfrak{H}$ is a factor group of $\mathfrak{I}(\mathfrak{o})/\mathfrak{r}(\mathfrak{o})$. Hence

Corollary 4.1 *Each ideal class (in the ordinary sense) of F contains infinitely many prime ideals.*

We shall see in the next section that for $F = \mathbf{Q}$, the above theorem gives us the classical theorem of Dirichlet on the prime numbers in an arithmetic progression. We also note that the theorem can be refined as follows. Let $\pi(x; c)$ denote the number of prime ideals $\mathfrak{p}$ in the coset c with $c \in \mathfrak{I}(\mathfrak{m})/\mathfrak{r}(\mathfrak{m})$ such that $N(\mathfrak{p}) \leq x$ $(x \geq 2)$. Then

$$\pi(x; c) = \frac{1}{N} \int_2^x \frac{du}{\log u} + O(xe^{-\alpha\sqrt{\frac{\log x}{n}}}),$$

which a constant $\alpha > 0$. Let $\pi(x)$ denote the number of prime ideals $\mathfrak{p}$ of F such that $N(\mathfrak{p}) \leq x$ $(x \geq 2)$. Then we obtain from the above that

$$\pi(x) = \int_0^x \frac{du}{\log u} + O(xe^{-\alpha\sqrt{\frac{\log x}{n}}}).$$

From this, we can deduce that

$$\lim_{x \to \infty} \frac{\pi(x)}{\frac{x}{\log x}} = 1.$$

4.3 Dirichlet's L-functions

In this section, we consider the case $F = \mathbf{Q}$. We then have $n = 1$, $r_1 = 1$, $r_2 = 0$, and $\sigma = \sigma_1 : J_\infty \to \mathbf{R}^*$, $J_\infty^0 = \sigma^{-1}(\mathbf{R}^+) = T$. Hence $X = X'$. Every ideal of $\mathbf{Q}$ is principal; given any $\mathfrak{a}$ in $\mathfrak{I}$, there exists a unique element $a \in \mathbf{Q}$, $a > 0$, such that $\mathfrak{a} = (a)$. The map $\mathfrak{a} = (a) \mapsto a$ defines an isomorphism

$$\mathfrak{I} \longrightarrow \mathbf{Q}^+ = \text{ the multiplicative group of positive rationals.}$$

The integral ideals in $\mathfrak{I}$ correspond to positive integers.

Let $\mathfrak{m}$ be such an integral ideal of $\mathbf{Q}$, and let $\mathfrak{m} = (m)$, $m \in \mathbf{Z}$, $m \geq 1$. We shall write $\mathfrak{I}(m)$, $\mathfrak{r}(m)$, etc. for $\mathfrak{I}(\mathfrak{m})$, $\mathfrak{r}(\mathfrak{m})$ etc. Let $m = \prod_{j=1}^{g} p_j^{t_j}$, $t_j \geq 1$ (p_j: distinct primes). Then, by the definition,

$$F(m) = \{a \mid a \in \mathbf{Q}, a > 0, \nu_{P_j}(a-1) \geq t_j, 1 \leq j \leq g\},$$

and $\mathfrak{r}(m) \leftrightarrow F(m)$ under $\mathfrak{I} \leftrightarrow \mathbf{Q}^+$.

Let

$$G(m) = (\mathbf{Z}/m\mathbf{Z})^*;$$

$G(m)$ is the multiplicative group of residue classes $\bar{u}$ $(= u \bmod \mathfrak{m})$ such that $(u, m) = 1$. It is an abelian group of order $\varphi(m)$ (φ = Euler's function). Let $u \in \mathbf{Z}$, $u \geq 1$, $(u, m) = 1$. Then the principal ideal (u) is in $\mathfrak{I}(m)$, and the coset of (u) mod $\mathfrak{r}(m)$ will be denoted by $c(u)$. Let $v \in \mathbf{Z}$, $v \geq 1$, $(v, m) = 1$. Then

$$c(uv) = c(u)c(v).$$

Also, $c(u) = c(v) \Leftrightarrow (\frac{u}{v}) \in \mathfrak{r}(m) \Leftrightarrow \frac{u}{v} \in F(m) \Leftrightarrow \nu_{P_j}(\frac{u}{v} - 1) \geq t_j$, $1 \leq j \leq g \Leftrightarrow u \equiv v \bmod p_j^{t_j}$, $1 \leq j \leq g \Leftrightarrow u \equiv v \bmod m$. Namely,

$$c(u) = c(v) \Longleftrightarrow u \equiv v \bmod m.$$

It follows that $\bar{u}(= u \bmod m) \mapsto c(u)$ defines an injective homomorphism of $G(m)$ into $\mathfrak{I}(m)/\mathfrak{r}(m)$. Let c be any element of $\mathfrak{I}(m)/\mathfrak{r}(m)$, and $\mathfrak{a} = (a)$, $a > 0$, an ideal in the coset c. Then $a = \frac{u}{v}$, $u, v \in \mathbf{Z}$, $u, v \geq 1$, $(u, m) = (v, m) = 1$. One sees immediately that $c = c(u)c(v)^{-1}$. Hence the above homomorphism is also surjective, and we obtain an isomorphism

$$G(m) \longrightarrow \mathfrak{I}(m)/\mathfrak{r}(m), \tag{4.4}$$

which maps $\bar{u}$ to $c(u)$ $(u \in \mathbf{Z}, u \geq 1, (u, m) = 1)$.

Let $w \in \mathbf{Z}$, $w \geq 1$, $(m, w) = 1$, and let $c = c(w) \in \mathfrak{I}(m)/\mathfrak{r}(m)$. Clearly prime ideals correspond to prime numbers under $\mathfrak{I} \to \mathbf{Q}^+$. Let $\mathfrak{p} = (p)$, $p =$ prime number. Then $\mathfrak{p} \in c = c(w) \Leftrightarrow c(p) = c(w) \Leftrightarrow p = w \bmod m$. Hence Theorem 4.1 implies that there exist infinitely many prime numbers p such that $p \equiv w \bmod m$. This is the classical theorem of Dirichlet.

Let $\Lambda(m)$ denote the group of all characters of $G(m)$; $\Lambda(m)$ is also an abelian group of order $\varphi(m)$. For $\lambda \in \Lambda(m)$, let $f_\lambda : \mathbf{Z} \to \mathbf{C}$ be defined by

$$f_\lambda(u) = \begin{cases} \lambda(\bar{u}), & \text{for } (u, m) = 1, \\ 0, & \text{otherwise.} \end{cases}$$

For simplicity, such a map f_λ will be also denoted by λ.

Let $(m_1) \subset (m_2)$, i.e. $m_2 \mid m_1$ $(m_1, m_2 \in \mathbf{Z}, m_1, m_2 \geq 1)$. Then the natural map $\mathbf{Z}/m_1\mathbf{Z} \to \mathbf{Z}/m_2\mathbf{Z}$ implies a surjective homomorphism $G(m_1) \to G(m_2)$, and hence also an injective homomorphism

$$\Lambda(m_2) \longrightarrow \Lambda(m_1).$$

A character λ in $\Lambda(m)$ is called primitive if λ is not in the image of $\Lambda(m') \to \Lambda(m)$ for any $m' \geq 1$, $m' \in \mathbf{Z}$, $m' \mid m$.

The isomorphism (4.4) defines a natural isomorphism $\mathrm{Ch}(\mathfrak{I}(m)/\mathfrak{r}(m)) \to \Lambda(m)$ so that we have

$$X(m) = X'(m) \longrightarrow \mathrm{Ch}(\mathfrak{I}(m)/\mathfrak{r}(m)) \longrightarrow \Lambda(m),$$

where the first isomorphism is defined in Sect. 4.1. Let $\chi \mapsto \psi_\chi^m \mapsto \lambda$ under the above maps. Then one sees easily that χ has the conductor (m) if and only if λ is primitive in $\Lambda(m)$. Suppose that this is so. Let $u \in \mathbf{Z}$, $u \geq 1$. If $(u, m) = 1$, then (u) is in $\mathfrak{I}(m)$, and $\psi_\chi((u)) = \psi_\chi^m((u)) = \lambda(\bar{u}) = \lambda(u)$ because $\bar{u} \leftrightarrow c(u) = (u) \bmod \mathfrak{r}(m)$ under $G(m) \to \mathfrak{I}(m)/\mathfrak{r}(m)$. On the other hand, if $(u, m) \neq 1$, then $\psi_\chi((u)) = 0$, $\lambda(u) = 0$, both from the definition of $\psi_\chi((u))$ and $\lambda(u)$. Hence

$$\psi_\chi((u)) = \lambda(u)$$

for any $u \in \mathbf{Z}$, $u \geq 1$. It follows that

$$\begin{aligned} L(s; \chi) &= \prod_{(p)} \left(1 - \psi_\chi((p)) N((p))^{-s}\right)^{-1} \\ &= \prod_{p} \left(1 - \lambda(p) p^{-s}\right)^{-1} \\ &= \sum_{u=1}^{\infty} \lambda(u) u^{-s}, \quad \mathrm{Re}\,(s) > 1. \end{aligned}$$

Therefore $L(s; \chi)$ is also denoted by $L(s; \lambda)$. We see that when λ ranges over all primitive characters in $\Lambda(m)$, $L(s; \lambda)$ gives us all L-functions $L(s; \chi)$ of $\mathbf{Q}$ for the characters χ in $X(m)$ with conductor (m). The L-functions of $\mathbf{Q}$ were first introduced by Dirichlet in this manner by means of the primitive characters of $G(m)$. Hence they are called Dirichlet's L-functions.

Let $\chi \mapsto \lambda$ as above, with λ primitive in $\Lambda(m)$. Let $\{m_1; s_1\}$ be the signature of χ. Since $\chi \in X' = X$, we have $s_1 = 0$ (cf. Sect. 4.1). We shall next show how to determine $e = m_1$ by means of λ. Let $\alpha \in J_\infty$, $\alpha = (1, \ldots, 1, x)$, $x \in \mathbf{R}^*$. Then

$$\chi(\alpha) = \left(\frac{x}{|x|}\right)^{m_1} |x|^{is_1} = \left(\frac{x}{|x|}\right)^{e}.$$

Put $x = -1$ so that $\alpha = (-1)_\infty$. Then $\chi((-1)_\infty) = (-1)^e$. Now, $\chi((-1)_0 \cdot (-1)_\infty) = \chi(-1) = 1$, $\chi((-1)_0 \cdot (m-1)_0) = \chi((1-m)_0) = 1$ because $(1-m)_0 \in U_0(m)$. Hence $\chi((-1)_\infty) = \chi((m-1)_0) = \psi_\chi((m-1)) = \lambda(m-1) = \lambda(-1)$. It follows that the signature of χ is $\{e; 0\}$ where

$$e = \begin{cases} 0, & \text{if } \lambda(-1) = 1, \\ 1, & \text{if } \lambda(-1) = -1. \end{cases}$$

In the above, we have implicitly assumed that $m > 1$. However, if $m = 1$, then $\chi \equiv 1$, $\lambda \equiv 1$ so that $e = 0$ and $\lambda(-1) = 1$. Hence the above formula holds also in this case.

With the above e, we then have

$$A(s; \chi) = \left(\frac{m}{\pi}\right)^{\frac{s+e}{2}}, \quad \gamma(s; \chi) = \Gamma\left(\frac{s+e}{2}\right),$$
$$\xi(s; \chi) = \left(\frac{m}{\pi}\right)^{\frac{s+e}{2}} \Gamma\left(\frac{s+e}{2}\right) L(s; \chi).$$

We can also prove that

$$W(\chi) = \frac{(-i)^e}{\sqrt{m}} \sum_{u=0}^{m-1} \lambda(u) e^{\frac{2\pi i u}{m}}.$$

Problem 4.2 Prove the above formula.

Let $m \in \mathbf{Z}$, $m \geq 1$. Consider

$$\prod_{\chi \in X(m)} L(s;\chi) = \prod_{\chi} \prod_{p} (1 - \psi_\chi((p))p^{-s})^{-1}$$

$$= \prod_{p} \left(\prod_{\chi} (1 - \psi_\chi((p))p^{-s}) \right)^{-1}, \quad \mathrm{Re}\,(s) > 1.$$

Suppose that $(p, m) = 1$. Then (p) is in $\mathfrak{I}(m)$, and $\psi_\chi((p)) = \psi_\chi^m((p)) = \lambda(\bar{p})$ if $\chi \mapsto \lambda$ under $X(m) \to \Lambda(m)$. Hence

$$\prod_{\chi \in \Lambda(m)} (1 - \psi_\chi((p))p^{-s}) = \prod_{\lambda \in \Lambda(m)} (1 - \lambda(\bar{p})p^{-s}).$$

Let f be the order of $\bar{p}$ in $G(m)$. Then f is the least positive integer such that $p^f \equiv 1 \bmod m$, and is a factor of $\varphi(m) = \#G(m)$. We put $\varphi(m) = fg$. Let $\eta = e^{\frac{2\pi i}{f}}$. Then it is well-known that when λ ranges over $\Lambda(m)$, $\lambda(\bar{p})$ ranges over $\{\eta^k \mid 0 \leq k < f\}$, g-times for each. Therefore the above product over $\Lambda(m)$ equals

$$\left(\prod_{k=0}^{f-1} (1 - \eta^k p^{-s}) \right)^g = (1 - p^{-fs})^g.$$

Let K be the cyclotomic field of mth roots of unity, i.e., $K = \mathbf{Q}(e^{\frac{2\pi i}{m}})$. Then the ideal (p) is decomposed in K into the product of prime ideals as follows:

$$(p) = \mathfrak{p}_1 \cdots \mathfrak{p}_g, \quad N(\mathfrak{p}_k) = p^f, \quad k = 1, \ldots, g;$$

here f and g are the integers defined in the above. It follows that

$$\prod_{\mathfrak{p} | p} (1 - N(\mathfrak{p})^{-s}) = \prod_{k=1}^{g} (1 - N(\mathfrak{p}_k)^{-s})$$

$$= (1 - p^{-fs})^g$$

so that

$$\prod_{\chi \in X(m)} (1 - \psi_\chi((p))p^{-s}) = \prod_{\mathfrak{p} | p} (1 - N(\mathfrak{p})^{-s}).$$

This is proved for any prime number p which is prime to m. However, we can obtain the same formula also for p which divides m. (The proof is similar, but is slightly more complicated). Therefore

$$\begin{aligned}\prod_{\chi\in X(m)} L(s;\chi) &= \prod_{p}\prod_{\mathfrak{p}|p}\left(1-N(\mathfrak{p})^{-s}\right)^{-1}\\ &= \prod_{\mathfrak{p}}\left(1-N(\mathfrak{p})^{-s}\right)^{-1}\\ &= \zeta_K(s), \quad \mathrm{Re}\,(s)>1,\end{aligned}$$

and we have the following

Theorem 4.2 *Let K be the cyclotomic field of m-th roots of unity, $m\geq 1$. Then*

$$\zeta_K(s) = \prod_{\chi\in X(m)} L(s;\chi), \quad s\in\mathbf{C}.$$

The product is taken our all Dirichlet's L-functions with characters χ in $X(m)$.

The above theorem has many interesting consequences. We know that both $\zeta_K(s)$ and $L(s;1)=\zeta_{\mathbf{Q}}(s)$ have a simple pole at $s=1$ and that $L(s;\chi)$, $\chi\not\equiv 1$, is holomorphic at $s=1$. Hence it follows from the above that $L(1;\chi)\neq 0$ for $\chi\not\equiv 1$ (cf. Sect. 4.2); for, otherwise, the zero of $L(s;\chi)$ at $s=1$ would cancel the pole of $L(s;1)$ at $s=1$, and $\zeta_K(s)$ would be holomorphic at $s=1$. Furthermore, since the residue of $\zeta_{\mathbf{Q}}(s)$ at $s=1$ is 1, it also follows from the above that the residue of $\zeta_K(s)$ at $s=1$ is equal to

$$\prod_{\substack{\chi\in X(m)\\ \chi\not\equiv 1}} L(1;\chi).$$

Hence (cf. Sect. 3.6) we have

$$h = \frac{w\sqrt{d}}{2^{r_1+r_2}\pi^{r_2}R}\prod_{\substack{\chi\in X(m)\\ \chi\not\equiv 1}} L(1;\chi)$$

for the class number h of K. In the next section, we shall compute the value of $L(1;\chi)$ $(\chi\not\equiv 1)$, and obtain the classical class number formula for the cyclotomic field K (with m a prime number).

Actually, the equality in Theorem 4.2 is a special case of the following important general result in class field theory. Let F be any number field and let Y be any finite subgroup of the group X of Hecke characters of F. Since X' is the torsion subgroup of X, Y is contained in X'. As stated in Sect. 4.1, X' may be identified with the character group of G, the Galois group of the maximal abelian extension A over F. Let H be the closed subgroup of all σ in G such that $\chi(\sigma)=1$ for χ in Y, and let K be the field, $F\subset K\subset A$, corresponding to the subgroup H of G. Then K/F is a finite abelian extension with Galois group G/H. For such an extension K of F, we then have

$$\zeta_K(s) = \prod_{\chi\in Y} L(s;\chi), \quad s\in\mathbf{C}.$$

Note that Y may be considered as the character group of G/H so that $Y \cong G/H$ (not canonically) and $[K : F] = [G : H] =$ order of Y. In Theorem 4.2, we have $F = \mathbf{Q}$ and $Y = X(m)$.

4.4 The Class Number Formula for Cyclotomic Fields

We first make a remark on the regulator of a number field F.

Let $\varepsilon_1, \ldots, \varepsilon_{r-1}$ $(r = r_1 + r_2)$ be any $r - 1$ units of F. Define

$$R(\varepsilon_1, \ldots, \varepsilon_{r-1}) = \left| \det \begin{pmatrix} e_1 \log |\varepsilon_1^{(1)}| & \cdots & e_r \log |\varepsilon_1^{(r)}| \\ \vdots & \vdots & \vdots \\ e_1 \log |\varepsilon_{r-1}^{(1)}| & \cdots & e_r \log |\varepsilon_{r-1}^{(r)}| \\ \frac{e_1}{n} & \cdots & \frac{e_r}{n} \end{pmatrix} \right|,$$

where $e_k = 1$ or 2 accordding as $1 \le k \le r_1$ or $r_1 + 1 \le k \le r$. If ε_1, ..., ε_{r-1} form a basis of E/W, then $R = R(\varepsilon_1, \ldots, \varepsilon_{r-1})$ is the regulator of F. Therefore, in such a case, $R(\varepsilon_1, \ldots, \varepsilon_{r-1})$ is independent of ε_1, ..., ε_{r-1}. In general, for any units $\varepsilon_1, \ldots, \varepsilon_m$, $m \ge 1$, let $E(\varepsilon_1, \ldots, \varepsilon_m)$ denote the subgroup of E generated by ε_1, ..., ε_m. Then $R(\varepsilon_1, \ldots, \varepsilon_{r-1}) \ne 0$ if and only if $E(\varepsilon_1, \ldots, \varepsilon_{r-1})$ (or $E(\varepsilon_1, \ldots, \varepsilon_{r-1})W$) has a finite index in E, and in that case, we have

$$R(\varepsilon_1, \ldots, \varepsilon_{r-1})/R = \left[E : E(\varepsilon_1, \ldots, \varepsilon_{r-1})W\right].$$

Now, let l be a fixed prime number, $l > 2$, and let K denote the cyclotomic field of l-th roots of unity;

$$K = \mathbf{Q}(\zeta), \quad \zeta = e^{\frac{2\pi i}{l}}.$$

Let

$$K_0 = \mathbf{Q}(\zeta + \overline{\zeta});$$

K_0 is the maximal real subfield of K contained in K. We put

$$\begin{aligned} n &= [K : \mathbf{Q}] = \varphi(l) = l - 1, \\ n_0 &= [K_0 : \mathbf{Q}] = \frac{l-1}{2}, \\ h &= \text{class number of } K, \\ h_0 &= \text{class number of } K_0, \\ R &= \text{regulator of } K, \\ R_0 &= \text{regulator of } K_0, \\ d &= |\text{discriminant of } K| = l^{n-1}, \end{aligned}$$

$$
\begin{aligned}
d_0 &= |\text{discriminant of } K_0| = l^{n_0-1},\\
E &= \text{group of units in } K,\\
E_0 &= \text{group of units in } K_0,\\
W &= \text{group of roots of unity in } K,\\
W_0 &= \text{group of roots of unity in } K_0.
\end{aligned}
$$

We can prove that $W = \{\pm\zeta^a\}, 0 \leq a < l$, so that $w =$ order of $W = 2l$. It is obvious that $W_0 = \{\pm 1\}$ and $w_0 =$ order of $W_0 = 2$. Let

$$
\begin{aligned}
g &= (2\pi)^{n_0} R/2l\sqrt{l}^{n-1} \left(= 2^{r_1+r_2}\pi^{r_2} R/w\sqrt{d} \text{ with } r_1 = 0, r_2 = n_0\right)\\
g_0 &= 2^{n_0} R_0/2\sqrt{l}^{n_0-1} \left(= 2^{r_1+r_2}\pi^{r_2} R_0/w_0\sqrt{d_0} \text{ with } r_1 = n_0, r_2 = 0\right)
\end{aligned}
$$

so that gh (resp. $g_0 h_0$) is the residue of $\zeta_K(s)$ (resp. $\zeta_{K_0}(s)$) at $s = 1$.

It is known that h_0 is a factor of h: $h = h_1 h_0$, $h_1 \in \mathbf{Z}$. Sometimes h_1 is called the first factor of the class number h, and h_0 the second factor of h. It can also be proved that

$$E = \{\zeta^a\} \times E_0.$$

It follows in particular that if $\varepsilon_1, \ldots, \varepsilon_{n_0-1}$ ($r = r_1 + r_2 = n_0$ for both K and K_0) form a basis of E_0/W_0, then they also form a basis of E/W. We then see immediately from the definition of regulators that

$$R = 2^{n_0-1} R_0$$

so that

$$g/g_0 = \pi^{n_0} 2^{n_0-1}/l\sqrt{l}^{n_0}.$$

For each integer a, prime to l, there exists a unique Galois automorphism σ_a of $K/\mathbf{Q}$ such that

$$\sigma_a(\zeta) = \zeta^a.$$

Clearly σ_a depends only upon the residue class $\overline{a}$ of a mod l. The map $\overline{a} \mapsto \sigma_a$ then defines an isomorphism

$$G(l) \cong G = \text{Galois group of } K/\mathbf{Q}.$$

The Galois group of K/K_0 corresponds in this isomorphism to the subgroup $\{\pm\overline{1}\}$. Hence

$$G(l)/\{\pm\bar{1}\} \cong G_0 = \text{Galois group of } K_0/\mathbf{Q}.$$

We fix an integer t (a primitive root mod l) such that $\bar{t}$ is a generator of $G(l)$, and we put $\sigma = \sigma_t$ so that $G = \{\sigma^k\}$, $0 \le k < n$. For any α in K, we then denote the conjugates of α by

$$\alpha^{(k)} = \sigma^k(\alpha), \quad 0 \le k < n = l - 1.$$

Let $\mathfrak{l}$ be the principal ideal of K generated by $1 - \zeta$:

$$\mathfrak{l} = (1 - \zeta).$$

Then $\mathfrak{l}$ is a prime ideal of K, and $(l) = \mathfrak{l}^n$. Hence $\sigma(\mathfrak{l}) = \mathfrak{l}$, and $\sigma(1-\zeta)/(1-\zeta) = (1-\zeta^t)/(1-\zeta)$ is a unit of K. Let

$$\eta = \zeta^{\frac{1-t}{2}}(1-\zeta^t)/(1-\zeta) = (\zeta^{\frac{t}{2}} - \zeta^{-\frac{t}{2}})/(\zeta^{\frac{1}{2}} - \zeta^{-\frac{1}{2}}).$$

(Note that $\zeta^{\frac{1}{2}}$ is in W because ζ has order l, $(l, 2) = 1$). It follows from the above that η is a unit, in fact, a real unit: $\eta \in E_0$. It is called the circular unit of K (for fixed t).

Since $G(l) = (\mathbf{Z}/l\mathbf{Z})^*$ is a cyclic group of order $n = l - 1$, so is the character group $\Lambda(l)$ (and also $X'(l) = X(l)$). Hence $\Lambda(l)$ has a unique character ρ of order 2: $\rho^2 \equiv 1$, $\rho \not\equiv 1$. Since $a \mapsto \left(\frac{a}{l}\right)$, $a \in \mathbf{Z}$, is such a character, we have

$$\rho(a) = \left(\frac{a}{l}\right), \quad a \in \mathbf{Z}.$$

Clearly ρ takes only real values (namely, 0, ± 1). However, if conversely λ is a character in $\Lambda(l)$ taking only real values, then $\lambda(a) = \pm 1$ for $(a, l) = 1$. Hence $\lambda^2 \equiv 1$, and we see that either $\lambda \equiv 1$ or $\lambda = \rho$ in the above. By this reason, we call ρ the real character in $\Lambda(l)$.

Let $\Lambda_0(l)$ denote the set of all λ in $\Lambda(l)$ such that $\lambda(-1) = 1$. Then $\Lambda_0(l)$ is a subgroup of order $n_0 = \frac{l-1}{2}$ in $\Lambda(l)$, and it may be considered as the group of all characters of $G(l)/\{\pm\bar{1}\}$. Since $\rho(-1) = \left(\frac{-1}{l}\right) = (-1)^{\frac{l-1}{2}}$, we see that $\rho \in \Lambda_0(l)$ or $\rho \notin \Lambda_0(l)$ according as $l \equiv 1 \bmod 4$ or $l \equiv 3 \bmod 4$.

Let

$$\omega(a) = \zeta^a = e^{\frac{2\pi i a}{l}}, \quad a \in \mathbf{Z}.$$

Then $\omega(a)$ depends only upon the coset $\bar{a}$ of a mod l, and it defines a character of the additive group of the finite ring (field) $\mathfrak{R} = \mathbf{Z}/l\mathbf{Z}$. For any λ in $\Lambda(l)$, we put

$$\nu(\lambda) = \sum_a \omega(a)\lambda(a) = \sum_{a=0}^{l-1} \lambda(a)\zeta^a.$$

$\nu(\lambda)$ is nothing but the Gaussian sum $G(\omega, \lambda)$ on $\mathfrak{R}$ defined in Sect. 2.3, and we see immediately from the lemmas proved there that

$$\sum_a \lambda(a)\zeta^{ua} = \bar{\lambda}(u)\nu(\lambda), \quad (u, l) = 1 \text{ or } \lambda \not\equiv 1, \tag{4.5}$$

$$\nu(\lambda) = 0, \quad \lambda = 1,$$
$$|\nu(\lambda)| = \sqrt{l}, \quad \lambda \not\equiv 1.$$

It follows that

$$\overline{\nu(\lambda)} = \sum_a \bar{\lambda}(a)\zeta^{-a} = \lambda(-1)\nu(\bar{\lambda}). \tag{4.6}$$

In particular, $\overline{\nu(\rho)} = \rho(-1)\nu(\rho)$. Hence $\nu(\rho)^2 = \rho(-1)|\nu(\rho)|^2 = \rho(-1) \cdot l = \pm l$, and $\nu(\rho) = \pm\sqrt{l}$ or $\pm i\sqrt{l}$ according as $l \equiv 1 \bmod 4$ or $l \equiv 3 \bmod 4$. It is not quite easy to determine the sign $\pm$ before $\sqrt{l}$ and $i\sqrt{l}$. One can show however that

$$\nu(\rho) = \begin{cases} \sqrt{l}, & l \equiv 1 \bmod 4, \\ i\sqrt{l}, & l \equiv 3 \bmod 4, \end{cases} \quad (\sqrt{l} > 0).$$

We know from Sect. 4.4 that

$$gh = \prod_{\substack{\chi \in X(l) \\ \chi \not\equiv 1}} L(1; \chi) = \prod_{\substack{\lambda \in \Lambda(l) \\ \lambda \not\equiv 1}} L(1; \lambda).$$

Applying a similar argument to K_0 (or, from the general theorem stated at the end of Sect. 4.4), we also see that

$$g_0 h_0 = \prod_{\substack{\lambda \in \Lambda_0(l) \\ \lambda \not\equiv 1}} L(1; \lambda).$$

To compute the value of $L(1; \lambda)$ for $\lambda \not\equiv 1$, we first prove the following,

Lemma 4.4 *Let ζ be any root of unity, $\zeta \neq 1$ (not necessarily $\zeta = e^{\frac{2\pi i}{l}}$). Let*

$$f(s) = \sum_{m=1}^{\infty} \zeta^m m^{-s}, \quad s > 0.$$

For any $\varepsilon > 0$, the above series converges uniformly for $s \geq \varepsilon$ so that $f(s)$ defines a continuous function of s for $s > 0$. In particular

$$\lim_{s \to 1} f(s) = f(1) = \sum_{m=1}^{\infty} \zeta^m m^{-1} = -\log(1 - \zeta),$$

where

$$\log(1-\zeta) = \log|1-\zeta| + i\arg(1-\zeta), \quad -\frac{\pi}{2} < \arg(1-\zeta) < \frac{\pi}{2}.$$

Proof Let $S(a) = \sum_{m=1}^{a} \zeta^m$ for $a \geq 1$, and $S(0) = 0$. If ζ is an N-th root of unity, $\zeta \neq 1$, then $\sum_{m=a}^{a+N} \zeta^m = 0$. Hence $S(a) = S(b)$ for $a \equiv b \bmod N$, and one obtains immediately that

$$|S(a)| \leq N, \quad a \in \mathbf{Z}.$$

Let $s \geq \varepsilon, b \geq a$. Then

$$\begin{aligned}\sum_{m=a}^{b} \zeta^m m^{-s} &= \sum_{m=a}^{b} (S(m) - S(m-1))\, m^{-s} \\ &= -S(a-1)a^{-s} + S(b)b^{-s} + \sum_{m=a}^{b-1} S(m)\left(m^{-s} - (m+1)^{-s}\right).\end{aligned}$$

Hence it follows from the above that

$$\begin{aligned}\left|\sum_{m=a}^{b} \zeta^m m^{-s}\right| &\leq Na^{-s} + Nb^{-s} + N\sum_{m=a}^{b-1} S(m)\left(m^{-s} - (m+1)^{-s}\right) \\ &= 2Na^{-s} \\ &\leq 2Na^{-s} \longrightarrow 0, \quad \text{for } a \longrightarrow +\infty\end{aligned}$$

This proves the uniform convergence. The fact that $\sum_{m=1}^{\infty} \zeta^m m^{-1} = -\log(1-\zeta)$ is well-known in the function theory.

Now, let $\lambda \not\equiv 1$, $s > 1$, and $\zeta = e^{\frac{2\pi i}{l}}$. Using

$$\frac{1}{l}\sum_{u=0}^{l-1} \zeta^{(a-m)u} = \begin{cases} 1, & m \equiv a \bmod l, \\ 0, & m \not\equiv a \bmod l, \end{cases}$$

one obtains that

$$\begin{aligned}L(s;\lambda) = \sum_{m=1}^{\infty} \lambda(m) m^{-s} &= \sum_{a=0}^{l-1} \lambda(a) \sum_{m \equiv a \bmod p} m^{-s} \\ &= \sum_{a=0}^{l-1} \lambda(a) \sum_{m=1}^{\infty} \frac{1}{l} \sum_{u=0}^{l-1} \zeta^{(a-m)u} m^{-s}\end{aligned}$$

$$= \frac{1}{l}\sum_{u=0}^{l-1}\sum_{a=0}^{l-1}\lambda(a)\zeta^{au}\sum_{m=1}^{\infty}\zeta^{-mu}m^{-s}$$

$$= \frac{\nu(\lambda)}{l}\sum_{u=0}^{l-1}\overline{\lambda}(u)\sum_{m=1}^{\infty}\zeta^{-mu}m^{-s} \qquad \text{(by (4.5))}$$

$$= \frac{\nu(\lambda)}{l}\sum_{u=1}^{l-1}\overline{\lambda}(u)\sum_{m=1}^{\infty}\zeta^{-mu}m^{-s} \qquad (\because \overline{\lambda}(0) = 0).$$

Since $\zeta^{-u} \neq 1$ for $1 \leq u \leq p-1$, we obtain from the lemma that

$$L(1;\lambda) = \lim_{s\to 1} L(s;\lambda) = -\frac{\nu(\lambda)}{l}\sum_{u=1}^{l-1}\overline{\lambda}(u)\log(1-\zeta^{-u})$$

$$= -\frac{\nu(\lambda)}{2l}\sum_{u=1}^{l-1}\overline{\lambda}(u)\left(\log(1-\zeta^{-u}) + \overline{\lambda}(-1)\log(1-\zeta^{u})\right).$$

Let $\lambda \in \Lambda_0(l)$ so that $\lambda(-1) = 1$. Let

$$\alpha = (1-\zeta)(1-\zeta^{-1}).$$

Then

$$L(1;\lambda) = -\frac{\nu(\lambda)}{2l}\sum_{u=1}^{l-1}\overline{\lambda}(u)\left(\log(1-\zeta^{-u}) + \log(1-\zeta^{u})\right)$$

$$= -\frac{\nu(\lambda)}{2l}\sum_{u=1}^{l-1}\overline{\lambda}(u)\log\sigma_u(\alpha)$$

$$= -\frac{\nu(\lambda)}{2l}\sum_{k=1}^{l-1}\overline{\lambda}(t)^k\log\alpha^{(k)},$$

where $\alpha^{(k)} = \sigma^k(\alpha) = \sigma_{t^k}(\alpha)$. Since α is contained in K_0, we have $\sigma^{n_0}(\alpha) = \alpha$, $\alpha^{(n_0+k)} = \alpha^{(k)}$. Also, $\lambda(t)^{n_0} = \lambda(t^{n_0}) = \lambda(-1) = 1$. Hence

$$L(1;\lambda) = -\frac{\nu(\lambda)}{l}\sum_{k=0}^{n_0-1}\overline{\lambda}(t)^k\log\alpha^{(k)}.$$

It follows that

$$h_0 = g_0^{-1}\prod_{\substack{\lambda\in\Lambda_0(l)\\ \lambda\neq 1}} L(1;\lambda) = |g_0^{-1}|\prod_{\lambda}{}' |L(1;\lambda)|$$

$$= \left(\frac{2^{n_0-1}R_0}{\sqrt{l}^{n_0-1}}\right)^{-1}\left(\frac{1}{\sqrt{l}}\right)^{n_0-1}\prod{}'\left|\sum_{k=0}^{n_0-1}\overline{\lambda}(t)^k\log\alpha^{(k)}\right|$$

$$= (2^{n_0-1}R_0)^{-1}\prod{}'\left|\sum_{k=0}^{n_0-1}\overline{\lambda}(t)^k\log\alpha^{(k)}\right|.$$

When λ ranges over the character group $\Lambda_0(l)$, $\overline{\lambda}(t)$ takes the values $e^{\frac{2\pi i}{n_0}m}$, $m = 0, 1, \ldots, n_0 - 1$. Hence by a well-known theorem on determinants, we have

$$\left|\prod_{\lambda\in\Lambda_0(l)}\sum_{k=0}^{n_0-1}\overline{\lambda}(t)^k\log\alpha^{(k)}\right| = \left|\det\begin{pmatrix}\log\alpha^{(0)} & \log\alpha^{(1)} & \cdots & \log\alpha^{(n_0-1)}\\ \log\alpha^{(1)} & \log\alpha^{(2)} & \cdots & \log\alpha^{(0)}\\ \vdots & \vdots & \ddots & \vdots\\ \log\alpha^{(n_0-1)} & \log\alpha^{(0)} & \cdots & \log\alpha^{(n_0-2)}\end{pmatrix}\right|.$$

Since $\eta = \zeta^{\frac{1-t}{2}}\frac{1-\zeta^t}{1-\zeta}$, $\eta^{(k)} = \zeta^{\frac{(1-t)t^k}{2}}\frac{1-\zeta^{t^{k+1}}}{1-\zeta^{t^k}}$,

$$|\eta^{(k)}|^2 = \frac{(1-\zeta^{t^{k+1}})(1-\zeta^{-t^{k+1}})}{(1-\zeta^{t^k})(1-\zeta^{-t^k})}$$

$$= \frac{\alpha^{(k+1)}}{\alpha^{(k)}}$$

Therefore the above determinant equals

$$\det\begin{pmatrix}2\log|\eta^{(0)}| & 2\log|\eta^{(1)}| & \cdots & 2\log|\eta^{(n_0-1)}|\\ 2\log|\eta^{(1)}| & 2\log|\eta^{(2)}| & \cdots & 2\log|\eta^{(0)}|\\ \vdots & \vdots & \ddots & \vdots\\ 2\log|\eta^{(n_0-2)}| & 2\log|\eta^{(n_0-1)}| & \cdots & 2\log|\eta^{(n_0-3)}|\\ \log\alpha^{(n_0-1)} & \log\alpha^{(0)} & \cdots & \log\alpha^{(n_0-2)}\end{pmatrix}.$$

Using $\sum_{k=0}^{n_0-1}\log|\eta^{(k)}| = \log N_{K_0/\mathbf{Q}}(\eta) = \log 1 = 0$ (because η is a unit), we then see that the determinant is again equal to the following:

$$\pm 2^{n_0-1}\left(\sum_{k=0}^{n_0-1}\log\alpha^{(k)}\right)\det\begin{pmatrix}0 & \log|\eta^{(1)}| & \cdots & \log|\eta^{(n_0-1)}|\\ 0 & \log|\eta^{(2)}| & \cdots & \log|\eta^{(0)}|\\ \vdots & \vdots & \ddots & \vdots\\ 0 & \log|\eta^{(n_0-1)}| & \cdots & \log|\eta^{(n_0-3)}|\\ 1 & \log\alpha^{(0)} & \cdots & \log\alpha^{(n_0-2)}\end{pmatrix}$$

$$= \pm 2^{n_0-1}\left(\sum_{k=0}^{n_0-1}\log\alpha^{(k)}\right)\det\begin{pmatrix} 0 & \log|\eta^{(1)}| & \cdots & \log|\eta^{(n_0-1)}| \\ 0 & \log|\eta^{(2)}| & \cdots & \log|\eta^{(0)}| \\ \vdots & \vdots & \ddots & \vdots \\ 0 & \log|\eta^{(n_0-1)}| & \cdots & \log|\eta^{(n_0-3)}| \\ 1 & 0 & \cdots & 0 \end{pmatrix}$$

$$= \pm 2^{n_0-1}\left(\sum_{k=0}^{n_0-1}\log\alpha^{(k)}\right)\det\begin{pmatrix} \log|\eta^{(0)}| & \log|\eta^{(1)}| & \cdots & \log|\eta^{(n_0-1)}| \\ \log|\eta^{(1)}| & \log|\eta^{(2)}| & \cdots & \log|\eta^{(0)}| \\ \vdots & \vdots & \ddots & \vdots \\ \log|\eta^{(n_0-2)}| & \log|\eta^{(n_0-1)}| & \cdots & \log|\eta^{(n_0-3)}| \\ \frac{1}{n_0} & \frac{1}{n_0} & \cdots & \frac{1}{n_0} \end{pmatrix}$$

$$= \pm 2^{n_0-1}\left(\sum_{k=0}^{n_0-1}\log\alpha^{(k)}\right) R(\eta^{(0)}, \ldots, \eta^{(n_0-2)}).$$

If $\lambda \equiv 1$, then

$$\begin{aligned} \sum_{k=0}^{n_0-1}\overline{\lambda}(t)^k\log\alpha^{(k)} &= \sum_{k=0}^{n_0-1}\log\alpha^{(k)} = \log N_{K_0/\mathbf{Q}}(\alpha) \\ &= \log N_{K/\mathbf{Q}}(1-\zeta) = \log p \\ &\neq 0. \end{aligned}$$

Hence we obtain from the above that

$$\prod_{\substack{\lambda\in\Lambda_0(l)\\ \lambda\not\equiv 1}}\left|\sum_{k=0}^{n_0-1}\overline{\lambda}(t)^k\log\alpha^{(k)}\right| = 2^{n_0-1}R(\eta^{(0)}, \ldots, \eta^{(n_0-2)}),$$

and consequently that

$$h_0 = \frac{R(\eta^{(0)}, \ldots, \eta^{(n_0-2)})}{R}.$$

Since $\eta^{(0)}\cdots\eta^{(n_0-1)} = N_{K_0/\mathbf{Q}}(\eta) = 1$, $E(\eta^{(0)}, \ldots, \eta^{(n_0-2)})$ is the subgroup of E_0 generated by all conjugates $\eta^{(k)}$ of η, and $E(\eta^{(0)}, \ldots, \eta^{(n_0-2)})W_0$ is the subgroup of E_0 generated by the conjugates $\pm\eta^{(k)}$ of $\pm\eta$. By the remark stated at the beginning of this section, we have the following formula for the class number h_0 of K_0:

$$h_0 = [E_0 : H];$$

here E_0 is the group of all units in K_0, and H is the subgroup of E_0 generated by the conjugates $\pm\eta^{(k)}$ of $\pm\eta$.

We shall next compute $h_1 = \frac{h}{h_0}$ from

$$h_1 = \frac{h}{h_0} = \frac{g_0}{g} \prod_{\substack{\lambda \in \Lambda(l) \\ \lambda \notin \Lambda_0(l)}} L(1; \lambda)$$
$$= \frac{l\sqrt{l}^{n_0}}{\pi^{n_0} 2^{n_0-1}} \prod_{\lambda} L(1; \lambda).$$

Let $\lambda \in \Lambda(l)$, $\lambda \notin \Lambda_0(l)$ so that $\lambda(-1) = -1$. Then

$$L(1; \lambda) = -\frac{\nu(\lambda)}{2l} \sum_{u=1}^{l-1} \overline{\lambda}(u) \left(\log(1 - \zeta^{-u}) - \log(1 - \zeta^{u})\right)$$
$$= -\frac{\nu(\lambda)}{2l} \sum_{u=1}^{l-1} \overline{\lambda}(u) \cdot 2i \arg(1 - \zeta^{-u}).$$

However,

$$\arg(1 - \zeta^{-u}) = \frac{1}{2}\left(2\pi - 2\pi \frac{u}{l}\right)$$
$$= \pi - \pi \frac{u}{l}.$$

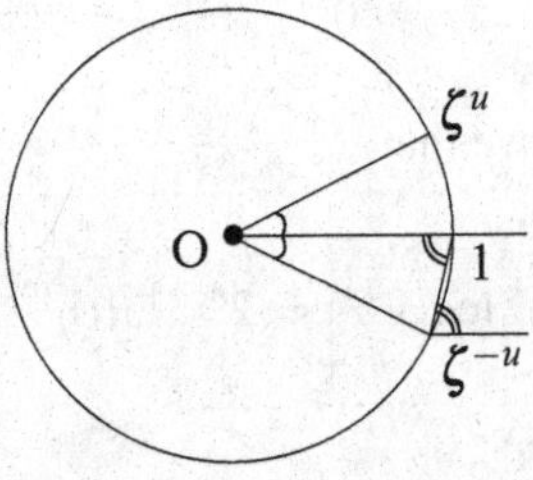

Since $\sum_{u=1}^{l-1} \overline{\lambda}(u) = 0$ for $\lambda \neq 1$ (or $\overline{\lambda} \neq 1$), we obtain

$$L(1; \lambda) = \frac{\pi i \nu(\lambda)}{l^2} \sum_{u=1}^{l-1} u \overline{\lambda}(u).$$

Therefore

$$h_1 = \frac{l\sqrt{l}^{n_0}}{\pi^{n_0} 2^{n_0-1}} \cdot \frac{\pi^{n_0} i^{n_0}}{l^{2n_0}} \cdot \prod_{\lambda} \nu(\lambda) \cdot \prod_{\lambda} \sum_{u=1}^{l-1} u \overline{\lambda}(u).$$

Let $l \equiv 1 \bmod 4$, $n_0 = \frac{l-1}{2} \equiv 0 \bmod 2$. Then $\rho \in \Lambda_0(l)$, and $\lambda \neq \overline{\lambda}$ for $\lambda \notin \Lambda_0(l)$. Hence the characters in the above product consist of $\frac{n_0}{2}$ pairs $\{\lambda, \overline{\lambda}\}$ with $\lambda \notin \Lambda_0(l)$. Since $\nu(\overline{\lambda}) = -\overline{\nu(\lambda)}$ by (4.6), we have $\nu(\lambda)\nu(\overline{\lambda}) = -|\nu(\lambda)|^2 = -l$. Hence

$$\prod_{\lambda} \nu(\lambda) = (-l)^{\frac{n_0}{2}} = i^{n_0}\sqrt{l}^{n_0}.$$

Let $l \equiv 3 \bmod 4$, $n_0 = \frac{l-1}{2} \equiv 1 \bmod 2$. Then $\rho \notin \Lambda_0(l)$, and the characters in the above product consist of the real character ρ and $\frac{n_0-1}{2}$ pairs $\{\lambda, \overline{\lambda}\}$. Since $\nu(\rho) = i\sqrt{l}$, we have

$$\prod_{\lambda} \nu(\lambda) = i\sqrt{l}(-l)^{\frac{n_0-1}{2}} = i^{n_0}\sqrt{l}^{n_0}.$$

Therefore, in either case,

$$\frac{l\sqrt{l}^{n_0}}{\pi^{n_0}2^{n_0-1}} \cdot \frac{\pi^{n_0} i^{n_0}}{l^{2n_0}} \cdot \prod_{\lambda} \nu(\lambda) = \frac{(-1)^{n_0}}{(2l)^{n_0-1}},$$

and we obtain the following formula for $h_1 = \frac{h}{h_0}$ (the first factor of the class number h of K):

$$h_1 = \frac{h}{h_0} = 2l \prod_{\substack{\lambda \in \Lambda(l) \\ \lambda(-1)=-1}} \left(-\frac{1}{2l} \sum_{u=1}^{l-1} u\lambda(u) \right).$$

By the same method, we can obtain similar class number formulas for more general types of cyclotomic fields as well as their subfields. For example, let $l \equiv 3 \bmod 4$, $l > 3$. Then $F = \mathbf{Q}(\sqrt{-l})$ is a subfield of K in the above. Let h' be the class number of F, and let

$$g' = \frac{\pi}{\sqrt{l}}.$$

Then

$$g'h' = L(1; \rho),$$

namely,

$$h' = \frac{\sqrt{l}}{\pi} L(1; \rho).$$

However, we have seen in the above that

$$L(1; \rho) = \frac{\pi i \nu(\rho)}{l^2} \sum_{u=1}^{l-1} u\rho(u) = \frac{\pi i (i\sqrt{l})}{l^2} \sum_{u=1}^{l-1} \left(\frac{u}{l}\right) u.$$

Hence

$$h' = -\frac{1}{l} \sum_{u=1}^{l-1} \left(\frac{u}{l}\right) u.$$

For $l = 3$, we have $g' = \frac{\pi}{3\sqrt{l}}$ so that we have to multiply the right-hand side of the above by 3. (We then obtain $h' = 1$).

Let $l \equiv 1 \bmod 4$. Then $\mathbf{Q}(\sqrt{l})$ is contained in the above cyclotomic field K, and we can obtain a class number formula for $\mathbf{Q}(\sqrt{l})$ similar to the formula $h_0 = [E_0 : H]$ for K_0.

Bibliography

1. E. Artin, The theory of algebraic numbers, Notes by Gerhald Würges from lectures held at the Mathematisches Institut, Göttingen, Germany in the Winter-Semester, 1956/57, Translated and distributed by George Sticker. Göttingen 172 pages, 1959.
2. E. Hecke, Vorlesungen über die Theorie der algebraischen Zahlen. Chelsea, 1948.

Reference added by Compiling Editors

3. J. Tate, Fourier analysis in number fields and Hecke's zeta-functions, in "Algebraic Number Theory" Edited by J. Cassels and A. Frohlich, pp. 305–347, Thompson Washington, D.C., 1967.

K. Iwasawa, *Hecke's L-functions*, SpringerBriefs in Mathematics,
https://doi.org/10.1007/978-981-13-9495-9

Printed in the United States
By Bookmasters